A-B Processes

A-B Processes

Towards Energy Self-sufficient Municipal Wastewater Treatment

Yu Liu, Jun Gu and Meng Zhang

IWA PUBLISHING

Published by **IWA Publishing**
Alliance House
12 Caxton Street
London SW1H 0QS, UK
Telephone: +44 (0)20 7654 5500
Fax: +44 (0)20 7654 5555
Email: publications@iwap.co.uk
Web: www.iwapublishing.com

First published 2019

British Library Cataloguing in Publication Data
A CIP catalogue record for this book is available from the British Library

ISBN: 9781789060072 (paperback)
ISBN: 9781789060089 (eBook)
ISBN: 9781789060096 (ePUB)

Contents

Preface

The conventional activated sludge (CAS) process has been applied for municipal wastewater treatment for more than 100 years, and is primarily based on the philosophy of biological oxidation by which organic matter is converted to biomass and carbon dioxide. However, the CAS process faces pressing challenges in terms of high energy consumption, generation of waste activated sludge (WAS) etc. Currently, almost all wastewater treatment plants (WWTPs) with CAS as a core process are being operated in an energy-negative fashion. It can be reasonably projected that there will be a substantial rise in energy demand with the more and more stringent effluent discharge standards being implemented worldwide. Obviously, high energy usage in current municipal WWTPs will seriously impact on the global wastewater industry, while is also indirectly connected to the issue of global climate change. Faced with such challenging situations with respect to demand for high quality product water, energy usage, handling of WAS, resources recovery, etc., we need to re-examine the current treatment philosophy by developing and adopting innovative process configurations and emerging technologies. In general, the potential chemical energy in municipal wastewater is almost five times that needed to drive the CAS process. Therefore, the solutions moving forward should rely on ways to enhance direct energy recovery from wastewater, while reducing in-plant energy consumption. In this sense, the A-B process – in which the A-stage is primarily designed for direct capturing or conversion of COD to methane gas via anaerobic treatment without producing excess sludge and the B-stage is designated for

Author(s): Yu Liu, Jun Gu and Meng Zhang
doi: 10.2166/9781789060089_xi

nitrogen removal – may offer a feasible engineering option for turning the operation of current municipal WWTPs from being energy-negative to energy self-sufficient.

Therefore, the five chapters in this book primarily focus on the main theme of environmental sustainability of WWTPs. Chapter 1 begins with a critical overview of the energy situation and challenges in current WWTPs, from which it appears that a shift of technology paradigm from the current "removal" philosophy to the concept of energy and resource recovery is inevitable. Chapter 2 presents a technology roadmap with the potential for realizing energy recovery and energy saving, together with possible resource recovery, towards the ultimate goal of energy self-sufficient municipal wastewater treatment. In Chapter 3, the concept of the A-B process is discussed in detail and various process configurations are presented in terms process performance, energy balance, potential challenges, etc. It appears that A-B process configurations are able to provide possible engineering solutions to make wastewater treatment energy self-sustainable. Chapter 4 further elucidates emerging biological processes, including mainstream deammonification, the integrated anaerobic membrane bioreactor, reverse osmosis and microalgal-bacterial processes, towards synergetic harvesting of water, energy and resources from municipal wastewater. A prospective outlook for A-B processes is further summarized in Chapter 5. It seems that the A-B process has the potential to be a game changer for the global market in municipal wastewater reclamation technology.

Finally, we would like to express our sincere thanks to Mark Hammond, Niall Cunniffe, Alan Peterson and IWA Publishing for all the support and assistance given to this book; meanwhile, we are also grateful for Yanqing Shen's assistance in editing of the materials. We hope that this book may be a useful reference for wastewater engineering professionals.

Yu Liu
Jun Gu
Meng Zhang
Nanyang Technological University
Singapore

About the Authors

Yu Liu. PhD, Professor of Environmental Engineering in the School of Civil and Environmental Engineering, Nanyang Technological University, Singapore. Dr Liu's research is primarily dedicated to innovative biological processes for wastewater and waste reclamation with special focus on energy and resources recovery.

Jun Gu. PhD, Research Fellow in Advanced Environmental Biotechnology Centre, Nanyang Environment & Water Research Institute, Nanyang Technological University, Singapore. Dr Gu's research mainly focuses on integrated processes for energy-efficient municipal wastewater reclamation.

Meng Zhang. PhD, Research Fellow in Advanced Environmental Biotechnology Centre, Nanyang Environment & Water Research Institute, Nanyang Technological University, Singapore. Dr Zhang's research focuses on biological processes for wastewater treatment, and energy and resource recovery from wastewater.

Contributors

Yu Liu
School of Civil and Environmental Engineering, Nanyang Technological University, Singapore

Jun Gu
Advanced Environmental Biotechnology Centre, Nanyang Environment & Water Research Institute, Nanyang Technological University, Singapore

Meng Zhang
Advanced Environmental Biotechnology Centre, Nanyang Environment & Water Research Institute, Nanyang Technological University, Singapore

Bin Ji
Department of Water and Wastewater Engineering, Wuhan University of Science and Technology, China

Hang Liu
Advanced Environmental Biotechnology Centre, Nanyang Environment & Water Research Institute, Nanyang Technological University, Singapore

Yingqun Ma
Advanced Environmental Biotechnology Centre, Nanyang Environment & Water Research Institute, Nanyang Technological University, Singapore

Abbreviations

AcoD	Anaerobic co-digestion
AD	Anaerobic digestion
AFBR	Anaerobic fixed bed reactor
Anammox	Anaerobic ammonia oxidation
AnAOB	Anaerobic ammonia oxidizing bacteria
AnMBR	Anaerobic membrane bioreactor
AOB	Ammonia oxidizing bacteria
BAF	Biological aerated filter
BES	Bio-electrochemical system
BOD	Biochemical oxygen demand
CANDO	Coupled aerobic-anoxic nitrous decomposition operation
CAS	Conventional activated sludge
CEPT	Chemically enhanced primary treatment
CHP	Combined heat and power
COD	Chemical oxygen demand
DBF	Denitrification biological filter
DEAMOX	Denitrifying ammonium oxidation
DHS	Down-flow hanging sponge
DO	Dissolved oxygen
EAM	Electrochemically active microorganisms
EDI	Electrodeionization
EGSB	Expanded granular sludge bed
FA	Free ammonia
F/M	Food to microorganisms
FNA	Free nitrous acid
GHG	Greenhouse gas

GWP	Global warming potential
HRAS	High rate activated sludge
HRAP	High rate algal pond
HRT	Hydraulic retention time
IE	Ion exchanger
IFAS	Integrated fixed-film activated sludge
MBBR	Moving bed biofilm reactor
MBR	Membrane bioreactor
MEC	Microbial electrolysis cell
MF	Microfiltration
MFC	Microbial fuel cell
MGD	Million gallons per day
MLE	Modified Ludzack-Ettinger
NAFO	Nitrate-dependent anaerobic ferrous oxidation
NAR	Nitrite accumulation rate
NaRR	Nitrate reduction rate
N-damo	Nitrite/nitrate-dependent methane oxidation
NER	Normalized energy recovery
NF	Nanofiltration
NiR	Nitrite reductase
NiRR	Nitrite reduction rate
NOB	Nitrite oxidizing bacteria
NRB	Nitrate respiring bacteria
pCOD	Particulate chemical oxygen demand
PDN	Partial denitrification
PN	Partial nitrification
RO	Reverse osmosis
SBR	Sequencing batch reactor
sCOD	Soluble chemical oxygen demand
SOFC	Solid oxide fuel cell
SOUR	Specific oxygen uptake rate
SRT	Sludge retention time
TN	Total nitrogen
TRL	Technology readiness level
TS	Total solids
TSS	Total suspended solids
UASB	Up-flow anaerobic sludge blanket
UF	Ultrafiltration.
VSS	Volatile suspended solids
VSR	Volatile solids reduction
WAS	Waste activated sludge
WRP	Water reclamation plant
WWTP	Wastewater treatment plant

Chapter 1

Energy use and challenges in current wastewater treatment plants

Meng Zhang and Yingqun Ma

1.1 INTRODUCTION

The world is facing an increasing energy crisis due to the crucial situation of growing energy requirements and rapid consumption of fossil fuels (Dudley, 2018). More than 2.8 billion (2.8×10^9) people worldwide still directly use solid fuels such as wood, coal and charcoal as energy sources instead of electricity (Drechsel *et al.*, 2015), and energy security is a critical issue in more and more countries across the world. In these circumstances, wastewater treatment has been regarded as a big energy drain. For example, in China, the volume of municipal wastewater produced had already reached 53.52 billion m^3 in 2015 (China MEP, 2017); supposing that $0.47\ kWh/m^3$ is required in the conventional activated sludge (CAS) process (Drechsel *et al.*, 2015; Zhou *et al.*, 2013), total energy consumption would therefore stand at $53.52 \times 0.47 = 25.15$ billion kWh. Similarly, it has been reported that the electrical energy consumed for municipal wastewater treatment in the United States accounted for 3% of the national electricity supply and 30–60% of municipalities energy demand (US EPA, 2006). Given the fact that a substantial amount of wastewater has not yet been treated properly across the world, it was estimated that a 44% increase in energy consumption for wastewater treatment could be expected between 2006 and 2030 (UNESCO, 2012). On the other hand, increasing energy usage is inevitably related to global climate change through intensified emissions of greenhouse

Author(s): Yu Liu, Jun Gu and Meng Zhang
doi: 10.2166/9781789060089_0001

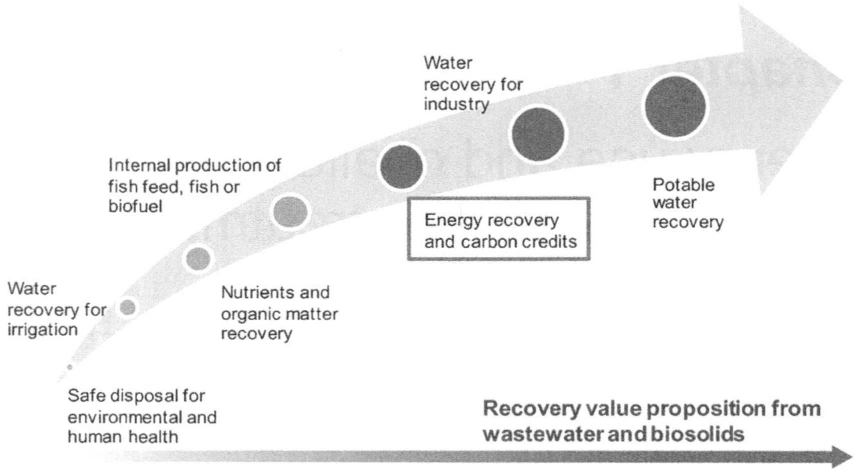

Figure 1.1 Value propositions for wastewater treatment considering investments and cost recovery potential. Adapted from Drechsel *et al.* (2015), with permission.

gases (GHGs). To keep global temperature increases below 2°C, the Paris Agreement on climate change put forward a central aim encouraging and promoting low-carbon energy industrial development (Armstrong *et al.*, 2016). Therefore, there is great interest in improving energy efficiency in wastewater treatment plants (WWTPs) with efforts being made to enhance energy recovery from wastewater.

The water–energy–climate nexus has gained increasing attention in recent years. Figure 1.1 shows value propositions for wastewater reclamation with energy recovery. Obviously, wastewater treatment is no longer only about pollution control but should be considered as a golden opportunity for water and energy recovery. To achieve an energy self-sufficient municipal wastewater treatment, this chapter thus attempts to offer a clear understanding of the energy situation and challenges in current WWTPs, including energy potential, energy consumption, energy recovery, GHG emissions and disposal of waste sludge.

1.2 ENERGY POTENTIAL IN MUNICIPAL WASTEWATER

Municipal wastewater is a unique source full of different substances containing a substantial amount of chemical energy. A better understanding of the energy potentially recoverable from municipal wastewater is critical to address the dilemma of the wastewater reclamation–energy nexus, while it is also essential for developing novel process configurations.

1.2.1 Theoretical energy potential in municipal wastewater

It is known that a large amount of chemical energy is associated with the organic matter and reactive nitrogen in municipal wastewater. The capture of organic matter, usually expressed in terms of chemical oxygen demand (COD) prior to biological oxidation, is helpful for maximizing the energy recovery potential from wastewater. Hence, a comprehensive analysis of the energy potential in municipal wastewaters is presented in this section.

1.2.1.1 Organic matter

Organic matter in wastewater has been widely explored as an important recoverable energy source. To capture chemical energy in wastewater organic matter, various anaerobic processes have been commonly employed in which biogas rich in methane (CH_4) is produced through anaerobic metabolisms of COD. The production of CH_4 from COD in anaerobic processes is mainly determined by methanogenesis, with a maximum conversion efficiency of 0.25 kg CH_4/kg COD, or 0.35 Nm^3 CH_4/kg COD (Equation 1.1) (Tchobanoglous *et al.*, 2003):

$$CH_4 + 2O_2 \rightarrow CO_2 + 2H_2O_{(l)} \quad \Delta H_R^{\ominus} = -890\,\text{kJ/mol } CH_4 \tag{1.1}$$

suggesting that 13.91 kJ of energy could be obtained from each gram COD removed from wastewater. For municipal wastewater with a typical COD concentration of 500 mg/L (Henze *et al.*, 2008), the recoverable energy can thus be estimated to be 500 $g/m^3 \times 13.91$ kJ/g = 6,955 kJ/m^3 wastewater.

1.2.1.2 Nitrogen

Energy can be recovered from two kinds of nitrogenous compounds during wastewater treatment: ammonia (NH_3) gas and nitrous oxide (N_2O) gas. According to Equations 1.2 and 1.3, the unit energy recoverable from NH_3 and N_2O gases are estimated to be 27.29 and 2.93 kJ/g N, respectively (Scherson *et al.*, 2013; Tchobanoglous *et al.*, 2003):

$$NH_3 + 0.75O_2 \rightarrow 0.5N_2 + 1.5H_2O \quad \Delta H_R^{\ominus} = -382\,\text{kJ/mol } NH_3 \tag{1.2}$$

$$N_2O \rightarrow N_2 + 0.5O_2 \quad \Delta H_R^{\ominus} = -82\,\text{kJ/mol } N_2O \tag{1.3}$$

For municipal wastewater with a typical ammonium-N concentration of 40 mg/L (Henze *et al.*, 2008), the recoverable energy from one cubic meter of wastewater can be estimated as 27.27 kJ/g N × 40 g N/m^3 = 1,092 kJ/m^3. If ammonia in municipal wastewater is directly captured as fertilizer for irrigation, this may partially offset the need for manufactured fertilizers. In the well-known Haber-Bosch process, the energy for ammonia production is about 69.48 kJ/g N (Gellings & Parmenter, 2016), thus the energy saving from direct use of wastewater ammonia can be roughly estimated to 69.48 kJ/g N × 40 g N/m^3 = 2,799 kJ/m^3 against the use of commercial ammonia-based fertilizer.

1.2.1.3 Phosphorus

Phosphorus is an essential and limited resource on earth and is usually regarded as a nutrient source rather than an energy source in municipal wastewater. It has been reported that the recoverable phosphorus in municipal wastewater could meet 35% of the global phosphorus fertilizer demand (Liu *et al.*, 2018). The energy associated with phosphorus processing has been estimated to be 7.60 kJ/g P (Gellings & Parmenter, 2016). Similar to the situation for ammonia, if wastewater phosphorus could be directly captured as fertilizer without "removal" through a highly complex biological process, the overall energy associated with phosphorus processing would be reduced substantially, i.e. $7.60\ kJ/g\ P \times 8\ g\ P/m^3 = 61\ kJ/m^3$, for municipal wastewater with a typical phosphorus-P concentration of 8 mg/L.

It should be noted that the constituents of municipal wastewater greatly depend on region, habit and climate. Table 1.1 summarises the average strength of energy-related constituents in municipal wastewater. Theoretically, the potential recoverable energy in a typical municipal wastewater falls into the range of 8047 kJ/m^3, i.e. 2.24 kWh/m^3, which is about five times the energy (averaged

Table 1.1 Energy potential in typical municipal wastewater.

Constituent[a]	Typical Concentration, g/m^3	Potential Recoverable Energy, kJ/m^3	Potential Energy Saving, kJ/m^3
Organic (COD)			
Total	500	6955[b]	
Biodegradable	320	4451	
Suspended	175	2434	
Dissolved	145	2017	
Inert	180	2504	
Suspended	80	1113	
Dissolved	100	1391	
Nitrogen	40	1092[c]	2799[d]
Organic	15	410	
Ammonia	25	682	
Phosphorus	8		61[d]
Total energy		8047	2860

[a]After Henze *et al.* (2008) and McCarty *et al.* (2011).
[b]Based on a theoretical energy of 13.91 kJ/g COD completely oxidized (Tchobanoglous *et al.*, 2003).
[c]Based on a theoretical energy of 27.29 kJ/g N (combustion with O_2 to N_2) (Scherson *et al.*, 2013; Tchobanoglous *et al.*, 2003).
[dd]Based on the energy required for producing N and P fertilizers, i.e. 69.48 kJ/g N by Haber-Bosch process and 7.60 kJ/g P (Gellings & Parmenter, 2016).

0.47 kWh/m^3) consumed in the CAS process (Wan *et al.*, 2016; Zhou *et al.*, 2013). These figures clearly suggest that municipal wastewater should not be regarded as a waste, but a source of energy which is more than enough to drive municipal wastewater treatment.

1.2.2 Potential energy sources: a global glance

In 2017, the primary energy consumption rate of the human population already stood at approximately 5.67×10^{14} MJ/year, of which 4.83×10^{14} MJ/year (i.e. about 85%) was generated from fossil fuel combustion (Dudley, 2018). Clearly, the overwhelming dependence on fossil fuels inevitably leads to a rapid rise of GHG emissions and depletion of non-renewable fuel stocks. In such circumstances, there is an urgent need to exploit alternative renewable energy sources to address the dilemma of energy shortage and global climate change in a more sustainable manner. According to REN21's report, renewable energy has shown positive developments in recent years, reaching a capacity of 33.94×10^{12} MJ/year in 2017 (Figure 1.2) (Sawin *et al.*, 2018).

As municipal wastewater contains a substantial amount of energy, it should be considered as an emerging renewable energy source. At a global scale, about 330–390 km^3/year of municipal wastewater (averaging at 360 km^3/year) is produced according to a model simulation and historical records (Flörke *et al.*, 2013; Sato *et al.*, 2013). Therefore, the total recoverable potential energy in municipal wastewater could be thermodynamically estimated as 8,047 kJ/m^3 $\times$ 360 km^3/year $= 2.90 \times 10^{15}$ kJ/year at the global scale, equivalent to 805 billion

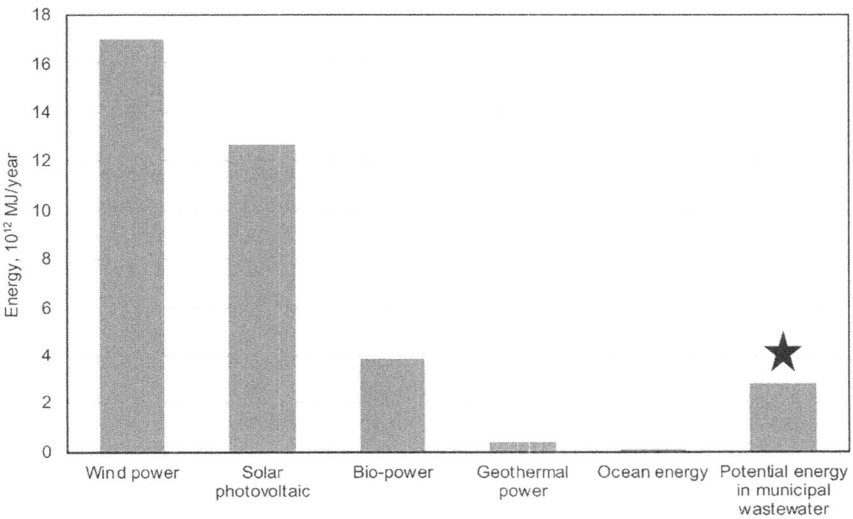

Figure 1.2 Global renewable energy capacity in 2017. Data from Sawin *et al.* (2018).

kWh of electricity per year. This implies that the potential energy in municipal wastewater is comparable to that from bio-power, accounting for about 8.53% of global renewable energy (Figure 1.2). Consequently, direct capture of the energy in municipal wastewater will have a definitely positive impact on reduced energy costs and GHG emissions.

1.2.3 Energy recovery from municipal wastewater in practice

In practice, part of the potential energy in municipal wastewater is unrecoverable during biological treatment, e.g. energy in organic matter oxidized to carbon oxide, the energy loss in the conversion of COD to methane, and energy required for microbial assimilation, usually accounting for 15% of the potential energy in municipal wastewater (McCarty *et al.*, 2011; Verstraete *et al.*, 2009). Unlike the current biooxidation-based treatment philosophy, if the anaerobic process is adopted as a lead, about 81% of the biodegradable COD could be directly converted to methane with minimized production of waste sludge. The methane gas produced needs to undergo a combined heat and power (CHP) process to maximize total energy recovery, with a typical CHP conversion efficiency of about 35%, which could be further increased to 40% by adopting innovative CHP technology, such as chemical fuel cells (US EPA, 2007). This is to say that about 32% (81% × 40%) of the potential energy in biodegradable wastewater COD could be harvested as electricity. Given the energy potential in biodegradable COD of 4451 kJ/m^3 (Table 1.1), the practically recoverable energy recovery from wastewater COD can be calculated to be 4451 $kJ/m^3 \times 32\% = 1{,}442$ kJ/m^3, i.e. 0.40 kWh/m^3. Recently, microbial fuel cells (MFC) have been proposed as an alternative for energy recovery from organic matter in wastewater. Limited by Coulombic loss and voltage efficiency, about 44% of potential energy in biodegradable COD can be recoverable as electricity, i.e. 4451 $kJ/m^3 \times 44\% = 1{,}959$ kJ/m^3, equivalent to 0.54 kWh/m^3.

Traditionally, nitrogenous compounds in municipal wastewater have been removed as unwanted pollutants, through various complex and energy-intensive biological nitrogen removal processes (Gu *et al.*, 2018; Hellinga *et al.*, 1998; Kartal *et al.*, 2010; Van Dongen *et al.*, 2001). To tackle such a challenging energy intensive situation, a change of mindset is required to explore possible nutrient recovery instead of simple removal, e.g. recovery and reuse of N and P for agricultural applications. Alternatively, some effort has been dedicated to direct energy recovery from nitrogen in the form of N_2O which is known as a powerful fuel additive. For this purpose, an innovative biological process called the Coupled Aerobic-anoxic Nitrous Decomposition Operation (CANDO) process had been developed (Scherson *et al.*, 2013). For a municipal wastewater with 40 g/m^3 of ammonium-N, if 80% of it could converted to N_2O, this could contribute an additional recoverable energy of 40 g $N/m^3 \times 80\% \times 2.93$ kJ/g $N = 93.76$

kJ/m^3, i.e. 0.03 kWh/m^3. Consequently, the practically recoverable energy in municipal wastewater may fall into the range of 0.43–0.57 kWh/m^3, which could likely offset the energy consumed in the CAS process. With the development of innovative process configurations or combinations, it could be feasible to achieve a more energy-efficient municipal wastewater treatment.

1.3 OVERVIEW OF ENERGY CONSUMPTION IN GLOBAL WWTPs

Manpower and energy demand constitute the major costs of most municipal WWTPs in the world. In recent years, in-plant energy consumption appears to be on a rising trend due to process upgrading and retrofitting in order to meet the tightened effluent discharge standards applied in more and more countries. For instance, in China, effluent discharge standards for municipal sewage treatment have been put forward to meet the Grade I-A standard countrywide, and even to meet the quasi Class IV standard of surface water in some provinces (Zhang *et al.*, 2016). According to a report by the US EPA, approximately 3% of the national electricity supply has been utilized to drive wastewater treatment, while up to 5% is used in other countries around the world (US EPA, 2006). Without any doubt, WWTPs are no longer negligible energy consumers globally.

1.3.1 Energy consumption in WWTPs

Energy consumption in WWTPs is largely determined by location, quality of influent wastewater, process configurations, treatment capacity and required effluent quality, etc. Typically, energy consumption for lagoons, trickling filters and activated sludge processes has been reported to be in the range of 0.09–0.29, 0.18–0.42 and 0.33–0.60 kWh/m^3, respectively (Logan, 2008). Today, the CAS process (and its variants) is most widely employed for large-scale municipal wastewater treatment with a capacity greater than 1 million gallons per day (Pabi *et al.*, 2013). In practice, energy demand is inversely related to the plant treatment capacity, with 10 million gallons per day as a cutoff above which further increases in capacity have limited effect on in-plant energy consumption.

Figure 1.3 presents the process layout of a typical system using the activated sludge process as the core technology for municipal wastewater treatment and in which the energy consumption in each unit/equipment is also indicated. An electricity input is mainly utilized to power the pumps, blowers, mixers, centrifuges, etc. The energy distribution in the activated sludge process is further shown in Figure 1.4, suggesting that aeration is the major energy consumer. In general, the energy used for aeration may account for almost 50% of the total in-plant energy demand, 16.0% for pumps, 11.9% for sludge thickening, 10% for anaerobic digestion, and about 10% for the other miscellaneous usages. For WWTPs with biological nitrogen removal, the aeration-associated energy

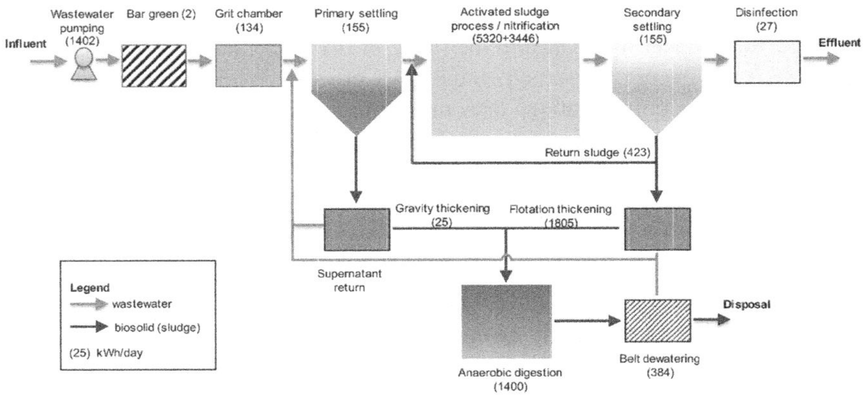

Figure 1.3 Illustration of a typical activated sludge process with the treatment capacity of 10 million gallons per day (numbers in bracket indicate the energy consumption by each unit for total COD removal and nitrification). Adapted from EPRI, 2002, with permission.

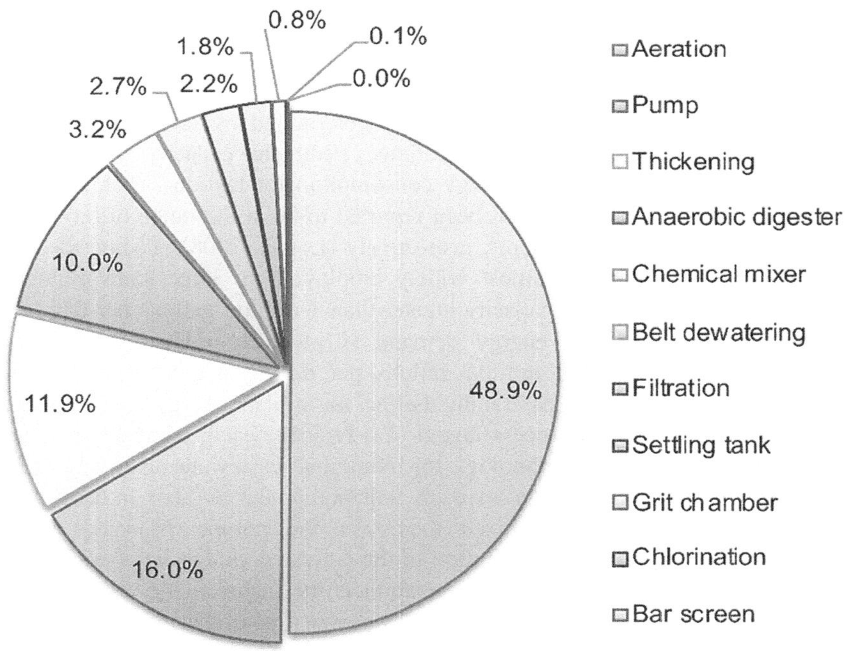

Figure 1.4 Distribution of energy consumption in an activated sludge process with a treatment capacity of 10 million gallons per day; data taken from Cao (2011).

consumption is often higher than that designed only for COD removal, with an additional energy consumption of about 0.1 kWh/m^3 (Goldstein & Smith, 2002; Monteith *et al.*, 2007). In some special circumstances, advanced treatment units, e.g. ultraviolet (UV), ozone (O_3), micro/nano/ultra-filtration (MF/NF/ UF), or even reverse osmosis (RO), may need to be adopted in municipal WWTPs for the purposes of water recycling and reuse. This definitely leads to an increase of the in-plant energy consumption.

1.3.2 Geographic distribution of energy consumption in WWTPs

Climate, economic level and living habits are geographically different across countries and definitely have a certain effect on the production and composition of the municipal wastewater produced. For example, the temperature in a tropical climate may favour the biological treatment of wastewater, resulting in a lower unit energy consumption. On the other hand, economic level may affect the choice of type of treatment process and equipment, while the living habits of the local population, to a great extent, determine the characteristics of the wastewater. Thus, it appears necessary to map the geographical distribution of energy consumption in WWTPs worldwide (Figure 1.5). As can be seen, the in-plant energy consumption by country was in a range of 0.26–0.68 kWh/m^3 with an average value of 0.47 kWh/m^3.

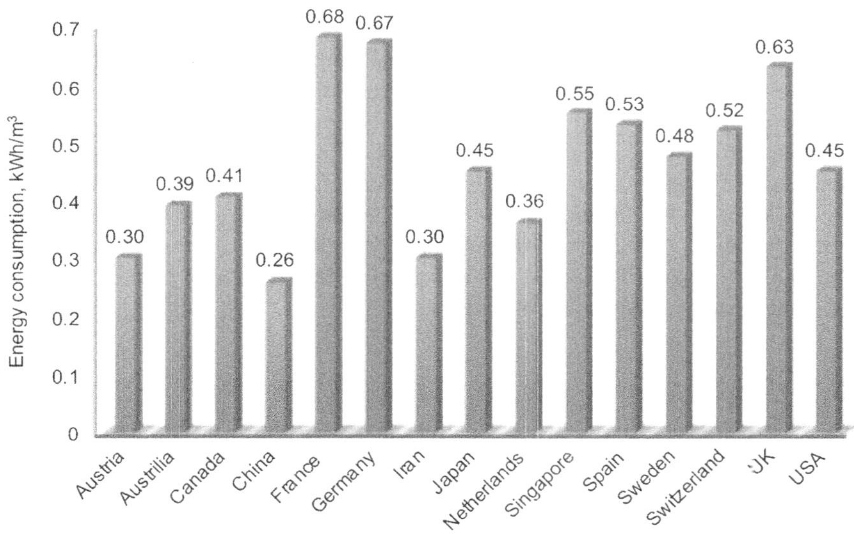

Figure 1.5 Geographic distribution of energy consumption in WWTPs. Data from Drechsel *et al.* (2015) and Zhou *et al.* (2013).

It should be noted that if kWh/m^3 is chosen as the unit indicator of in-plant energy consumption, it does not allow a cross comparison over different countries since wastewater strengths in terms of COD, nitrogen, phosphorus and TSS may be very different. One example is that the energy consumption of WWTPs in Austria and Sweden were reported to be 0.30 and 0.48 kWh/m^3, respectively; they are also reported using a population equivalent (pe) based expression with values of 23 and 42 kWh/(pe· COD· year), respectively, showing a more significant difference, likely due to the two countries' different living habits (Jonasson & Ulf Jeppsson, 2007). This indeed is supported by another example in Austria and Germany, where energy consumption measured in kWh/m^3 showed a two-fold difference, while they were found to be comparable when kWh/(pe· COD· year) was used as the indicator, with both reported at 23 kWh/(pe· COD· year) (Cao, 2011).

It is generally accepted that countries can be roughly clustered into developed and developing countries according to their economic development levels. In this regard, the salient points derived from Figure 1.5 are that no significant differences in the unit energy consumption for municipal wastewater treatment were observed across the developed countries, whereas relatively low energy consumption was noted in the developing countries due to many various reasons. For example, the unit energy consumption for wastewater treatment in China was as low as 0.26 kWh/m^3 and the annual electrical energy consumption associated with municipal wastewater treatment only accounted for 0.3% of the national electrical energy demand (Hao *et al.*, 2015), which was about 10 times lower than the situation in the USA. Generally, a low unit energy consumption as observed in Figure 1.5 may imply inadequate wastewater collection and treatment. With the more and more stringent effluent discharge standards being implemented in many developing countries, it can be reasonably projected that an increasing energy demand for municipal wastewater treatment is inevitable in the near future. According to the International Energy Agency, the energy demand for wastewater treatment will jump by about 44% from 2006 to 2030 (UNESCO, 2012). These factors all clearly point to the fact that an energy efficient wastewater treatment technology is urgently needed in order to timely address the challenging wastewater–energy–climate change nexus.

1.3.3 Global energy consumption of WWTPs

Global annual wastewater generation has been estimated to be 309.52 Gm^3/year (AQUASTAT, 2018), 45.5% of it generated by Asia followed by the Americas, Europe, Africa and Oceania (Table 1.2). On average, only about 60% of the produced municipal wastewater (184.90 Gm^3/year) had been treated. It appears that the generation of municipal wastewater was correlated to individual countries incomes. According to Sato *et al.* (2013), the averaged wastewater treatment

Table 1.2 Wastewater generation, treatment capacity and associated energy consumption in various regions of the world.

Region	Generated Wastewater[a], Gm^3/year	Treated Wastewater[a], Gm^3/year	Treated Ratio	Energy Consumption[b], GkWh/year	Electricity Consumption[c], GkWh/year	Proportion of Energy Consumption, %
Africa	14.01	6.88	0.49	3.23	663	0.49
Asia	140.80	85.44	0.61	40.16	10,754	0.37
Americas[d]	98.45	60.47	0.61	28.42	5,691	0.50
Europe	56.27	32.11	0.57	15.09	4,634	0.33
Oceania	2.09	2.28	1.09	1.07	273	0.39
World	309.52	184.90	0.60	86.90	22,015	0.39

[a]Data from AQUASTAT (2018).
[b]Data based on the unit energy consumption of 0.47 kWh/m^3.
[c]Data from Enerdata (2018).
[d]Data for North America and Latin America.

ratios in high-income, upper-middle-income, lower-middle-income and low-income countries were subjected to a decreasing trend from 70% to 8%.

If the averaged unit energy consumption for wastewater treatment is supposed to be 0.47 kWh/m^3, the energy consumed for municipal wastewater treatment could reach $0.47 \times 184.90 = 86.90$ GkWh/year. In fact, electricity consumption by wastewater treatment already accounted for 0.33–0.50% of global electricity demand in 2017.

1.4 CURRENT ENERGY RECOVERY IN WWTPs

To date, almost all WWTPs have been seriously challenged by their intensive energy consumption. The current biological processes for municipal wastewater treatment have been designed and operated according to the philosophy of bio-oxidation, by which roughly 40–50% of soluble COD in wastewater is oxidized to carbon dioxide, while the rest is synthesized into biomass known as waste sludge. Therefore, in WWTPs, anaerobic digestion (AD) of waste sludge becomes the sole means of energy recovery, in the form of biogas.

1.4.1 Anaerobic digestion of waste sludge for energy recovery

AD is a widely deployed technology for energy recovery from waste sludge in current WWTPs, by which the organic fraction of waste sludge can be converted to biogas with 50–70% volume of CH_4, a renewable energy fuel. Meanwhile, AD also leads to a reduction of sludge volume and to the destruction of harmful pathogens in the waste sludge (Appels *et al.*, 2008). During AD, waste sludge undergoes a series of biochemical reactions carried out by different functional microbial species, i.e. hydrolysis, acidogenesis, acetogenesis and methanogenesis, among which hydrolysis has been identified to be the step limiting the overall AD efficiency (Henze *et al.*, 2008).

As a highly complex biological process, AD requires strict anaerobic conditions, with an oxidation reduction potential (ORP) below 200 mV and various operational parameters – e.g. pH, alkalinity, temperature, sludge retention time (SRT) – all affecting the overall efficiency. Extensive efforts have been made to optimize the process performance with the aim of enhancing methane production. In general, pH and SRT for AD should be controlled in the ranges of 6.5–7.2 and 10–25 days, respectively, especially for methanogenic bacteria responsible for methane production, while the temperature needs be maintained at 28–40°C for mesophilic and 50–57°C for thermophilic AD.

A representative AD process for energy recovery from waste sludge in WWTPs is depicted in Figure 1.6. The waste sludge fed to an anaerobic digester usually contains both primary and secondary sludge. In general, the primary sludge

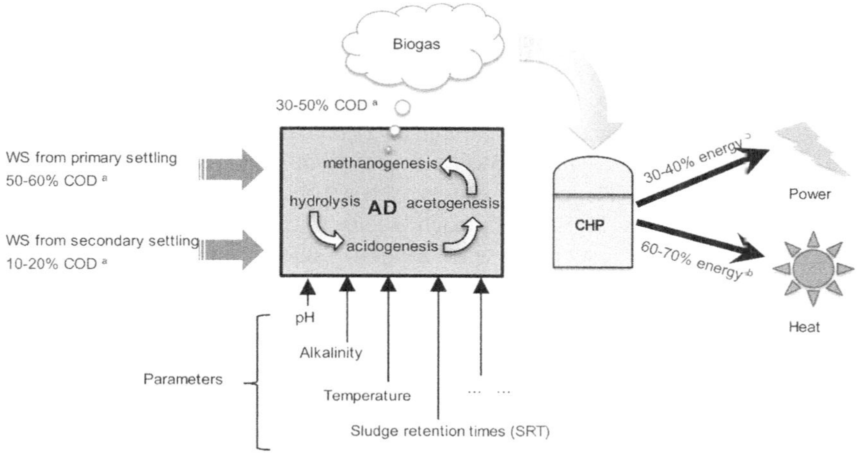

Figure 1.6 Illustration of a typical AD process for energy recovery from waste sludge. [a] Data from Wan *et al.* (2016); [b] Data from US EPA (2007) and McCarty *et al.* (2011). WS = waste sludge.

consists of 50–60% of suspended solids and 30–40% of biodegradable organics in terms of biochemical oxygen demand (BOD), while the secondary sludge originates from wasted biomass from the biological process, i.e. waste activated sludge (WAS). Currently, only about 30–50% of the influent wastewater COD equivalent can be converted to biogas through AD of waste sludge. The produced biogas, with methane as the main component, is further combusted through CHP to generate electricity (with an efficiency of 30–40%) and heat (with an efficiency rate of 60–70%). Municipal wastewater usually has a typical COD concentration of about 500 mg/L (Mihelcic *et al.*, 2014). Supposing 40% of influent COD is converted to methane COD via AD of waste sludge and the recoverable energy from methane COD is about 13.91 kJ/g, the unit recoverable energy via AD of waste sludge can therefore be estimated as 500 g/m^3 × 0.40 × 13.91 kJ/g = 2782 kJ/m^3. At an electricity conversion efficiency of 35%, a total electrical energy of about 973.7 kJ/m^3 (i.e. 0.27 kWh/m^3) would be ultimately produced via AD of waste sludge, meanwhile 1,808.3 kJ/m^3 would be harvested as heat from methane combustion. Obviously, this part of heat energy is not negligible and should be properly used in WWTPs.

1.4.2 Energy efficiency in current WWTPs

Currently, energy consumption in the CAS process is normally in a range of 0.26–0.68 kWh per m^3 of wastewater treated with an average of 0.47 kWh/m^3.

As discussed above, about 0.27 kWh/m^3 of electrical energy could be recovered through AD of waste sludge, indicating that the energy recovered in a current WWTP could offset about 50–60% of total input energy consumption.

Table 1.3 summarizes the energy situation in WWTPs located in different countries. It can be seen that the current energy efficiency in most WWTPs is in a range of 31–61%, which indeed is consistent with the estimated value, indicating the big challenge of energy self-sufficient municipal wastewater treatment. As such, a holistic approach needs to be adopted to reduce the energy consumptions associated with aeration, sludge thickening/dewatering and pumping in a synergistic manner. It should be pointed out that the different energy efficiencies observed in Table 1.3 may be attributed to differences in wastewater characteristics, treatment process, operation conditions, local climate, type of equipment etc. For example, the influent COD and BOD concentrations in the Strass WWTP were reported to be 547 and 291 mg/L, respectively (Jonasson & Ulf Jeppsson, 2007), while they were only 346 and 173 mg/L, respectively, for wastewater in the Gaobeidian WWTP (Gans *et al.*, 2007). It is commonly accepted that energy recovery potential is related to wastewater COD concentration. Alternatively, upgrading of system equipment may help to improve overall in-plant energy efficiency, e.g. the AD efficiency of the Strass WWTP was improved from 33% to 38% after installation of a new CHP unit.

The Strass WWTP is located in Austria with a treatment capacity of 6 million gallons per day, equivalent to 22,800 m^3/d, and its treatment process was designed to have two stages: the first stage being for COD removal with an HRT of 30 min and SRT of 12 h, and the second stage specifically designed for biological nitrogen removal with an SRT of 10 days. In the Strass WWTP, three operation strategies were successfully adopted to gradually improve energy efficiency from 49% to 100%. Firstly, the CHP unit was upgraded to enhance energy recovery from the combustion of bio-methane produced through AD of the waste sludge. With this measure in place, electrical efficiency rose from 33% to 38%, yielding a total power of 340 kW. Secondly, a deammonification process was implemented for sidestream nitrogen removal, with which energy consumption was reduced by 44%. Moreover, the deammonification of the sidestream did not need supplementary organic carbon which could be used for biogas production via AD, resulting in energy efficiency increasing from about 59% to 62%. Thirdly, an intermittent aeration device controlled by on-line effluent ammonia concentration was used to replace the conventional fine-bubble diffusers and the energy consumption associated with aeration was reduced substantially.

Although most WWTPs are currently being operated in an energy-negative fashion, there is still a great possibility and opportunity to turn current operation towards energy self-sufficiency through concurrently reducing in-plant energy consumption while improving energy recovery potential.

Table 1.3 Summary of specific energy consumption and energy efficiency in current WWTPs.

Energy Consumption[a], kWh/m^3	**Austria (Strass) (Jonasson & Ulf Jeppsson, 2007)**	**China (Gaobeidian) (Gans *et al.*, 2007)**	**Japan (Mizuta & Shimada, 2010)**	**Iran (Nouri *et al.*, 2006)**	**Singapore (Ulu Pandan) (Cao, 2011)**
Total energy consumption	0.317	0.258	0.320	0.300	0.440
Aeration	0.181	0.148	0.148	0.231	0.187
Sludge thickening and dewatering	0.040	0.005	0.100	0.021	0.025
Pumping	0.028	–	0.059	0.034	0.079
Total energy generation	0.346	0.081	0.170	0.182	0.150
Energy efficiency[b], %	109	31	50	61	34

[a]Total energy consumption and main energy consumers.
[b]Energy efficiency defined as the percentage of reduced and generated energy of total required energy of a WWTP.

1.5 THE WASTEWATER TREATMENT–ENERGY–GREENHOUSE GAS NEXUS IN CURRENT WWTPs

As discussed above, municipal wastewater treatment has become a non-negligible energy consumer and is speeding up the exhaustion of non-renewable fossil fuels, while producing substantial amounts of greenhouse gases (GHGs) including carbon dioxide, nitrous oxide and methane. This poses a serious question about the environmental sustainability of current WWTPs (Lee *et al.*, 2017). Indeed, as highlighted in the IPCC guidelines, WWTPs are listed as a contributor to GHG emissions (Eggleston *et al.*, 2006). Obviously, the energy consumed for wastewater treatment would have an adverse impact on the present global warming, e.g. approximately 1.14×10^8 tons of CO_2 is released annually in producing electricity required to treat wastewater in China (Hao *et al.*, 2015). On a related issue to carbon dioxide, the European Union's Emissions Trading Scheme was established about a decade ago, by which carbon dioxide can be traded as a commercially valuable carbon credit. Recently, the price of carbon dioxide dramatically jumped to about €20 per ton (Keating, 2018). These points clearly imply that the emission of carbon dioxide from WWTPs might be no longer free and associated fees should be counted towards overall operation costs in the future. For the purpose of illustration, the emission of 1.14×10^8 tons of CO_2 might result in an additional cost of €20/ton × (1.14×10^8 tons/year) = €2.28 billion/year, equivalent to about RMB 17.74 billion/year. This would, significantly, contribute an additional treatment cost of (RMB17.74 billion/year)/(54.4 billion m^3 of wastewater in 2015) = RMB 0.326/m^3. Moreover, Germany's Berenberg Bank has even predicted that the trading price of carbon dioxide could further rise to €100 per ton by 2020 (Keating, 2018). If the present wastewater treatment processes remain as "status quo", the hidden treatment costs associated with carbon dioxide emissions would likely become a non-negligible game changer in a future scenario of biological wastewater treatment.

1.5.1 Greenhouse gas emission pathways in WWTPs

The GHGs emitted from WWTPs are mainly carbon dioxide (CO_2), methane (CH_4) and nitrous oxide (N_2O), of which CH_4 and N_2O are potent GHGs with a 100-year global warming potential (GWP) of 298 times and 34 times higher than carbon dioxide, respectively (Myhre *et al.*, 2013). The concept of a carbon footprint has been widely recommended to quantify the total GHG emissions in WWTPs in terms of equivalent CO_2 (CO_2e). It should be noted that only the CO_2 from fossil fuel sources is included in the carbon footprint estimation, whereas generated CO_2 in WWTPs is mostly biogenic, which can be emitted to the atmosphere through a natural decay process anyway and thus should be excluded (Crawford *et al.*, 2011). As illustrated in Figure 1.7, the GHG generation in WWTPs may result from direct and indirect pathways. Direct GHG emission refers to GHG emitted from all wastewater treatment units, basically including any biological

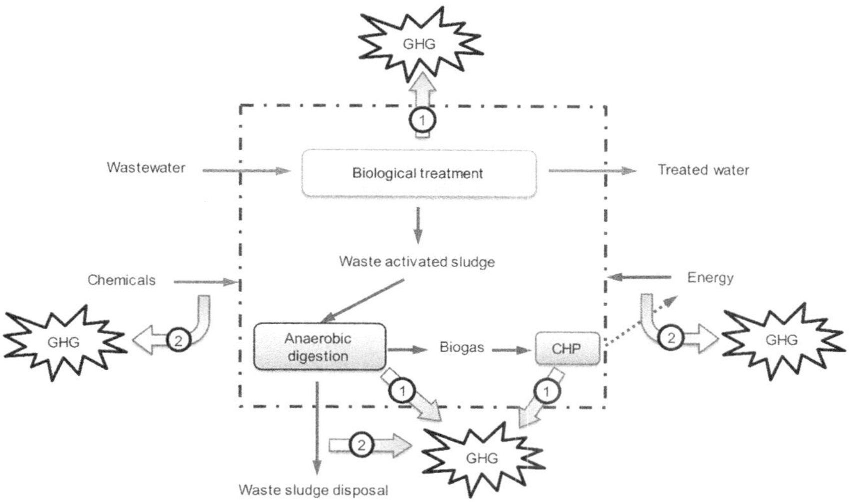

Figure 1.7 GHG emissions from WWTPs with anaerobic digestion. Dotted line refers to the WWTPs boundary. ① and ② refer to direct and indirect GHG emissions, respectively.

treatment unit and sludge treatment unit, while indirect pathways mainly refer to the GHG emission associated with direct and indirect energy consumption associated with wastewater treatment, e.g. energy utilized for wastewater treatment, chemical consumption, post disposal of waste sludge etc (Parravicini *et al.*, 2016).

1.5.1.1 Direct GHG emissions in WWTPs

CH_4 and N_2O are the main direct GHG emission from WWTPs. CH_4 generally originates from anaerobic metabolism of organic matters in wastewater. In fact, all anaerobic conditions may contribute to the emission of CH_4 in the presence of organic matter or biosolids (Monteith *et al.*, 2005). In a WWTP, the potential sources of CH_4 may include the primary settling tank with deep sludge blankets, the sludge thickening tank, anaerobic digesters, sludge storage tanks, or seepage from the combined heat and power process, etc. In WWTPs with anaerobic digestion for sludge treatment, CH_4 generation will be enhanced for energy recovery, and so are the CH_4 emissions. In this situation, the sludge line is supposed to be the main source of direct CH_4 emission, which can account for more than 75% of all the emitted CH_4 in WWTPs (Daelman *et al.*, 2012).

Biological nitrogen removal processes are the main sources of N_2O emission. In general, N_2O can be produced through three biological reactions, i.e. hydroxylamine (NH_2OH) oxidation, nitrifier denitrification, and denitrification (Schreiber *et al.*, 2012; Wunderlin *et al.*, 2012). To date, N_2O emission through the pathways of hydroxylamine oxidation and nitrifier denitrification have been commonly reported and are usually mediated by ammonia oxidizing bacteria (Ahn *et al.*, 2010; Sabba *et al.*, 2018). Moreover, N_2O has also been identified as an intermediate of denitrification by autotrophic or heterotrophic bacteria (Shapleigh, 2013). Many factors including dissolved oxygen concentration, nitrite concentration, and COD/N ratio, etc., may affect N_2O emission during municipal wastewater treatment (Massara *et al.*, 2017).

1.5.1.2 Indirect GHG emissions in WWTPs

Today, most electrical energy is still generated through the combustion of non-renewable fossil fuels, thus high energy utilization in municipal WWTPs may lead to a large indirect GHG emission. For example, accurate values of GWP for external energy production have been reported as 0.94 kg CO_2e/kWh, 0.58 kg CO_2e/kWh and 0.75 CO_2e/kWh in coal-, gas- and oil-burning power plants, respectively (Qi & Chang, 2013). In addition, the transportation, production and disposal of waste sludge would also involve consumption of fossil fuel, contributing to further GHG emissions.

1.5.2 Estimation of GHG emissions in WWTPs

Any estimation of GHG emissions and carbon footprint is highly dependent on emission factors (EF). The EFs for direct and indirect GHG emissions in WWTPs are summarized in Table 1.4. For the purpose of illustration, the Ulu Pandan wastewater reclamation plant (WRP) in Singapore is taken as a typical case to estimate GHG emissions. The Ulu Pandan WRP uses conventional activated sludge as the core process for wastewater treatment, with AD of waste sludge for energy recovery (Cao, 2011). In this WRP, influent wastewater was found to contain 650 mg/L of COD and total nitrogen (TN) of 55 mg N/L. At a steady state, 44.9% of the influent COD was fed into the anaerobic digesters, with 17.9% being converted to biogas, of which 64% was CH_4 gas. As shown in Table 1.3, total energy consumption and generation were estimated to be 0.44 and 0.15 kWh/m^3, respectively.

Given this situation, direct CH_4 emission can be estimated as 650 mg/L $\times$ 17.9% $\times$ 64% $\times$ 10.4% = 7.7 mg CH_4/L, i.e. 263 g CO_2e/m^3. The EF of N_2O is calculated to be 1.48% according to the relationship between TN removal and EF presented in Table 1.4. In this regard, the direct N_2O emission can be estimated as 55 mg N/L $\times$ 1.48% mg N_2O N/mg N = 0.81 mg N_2O/L, i.e. 243 g CO_2e/m^3. Therefore, the total direct GHG emission was estimated to be 506 g CO_2e/m^3. The external energy input was the difference between the energy consumed and energy recovered, i.e. 0.44−0.15 = 0.29 kWh/m^3. Thus, the indirect GHG emission associated with this part of electrical energy would stand at 0.29 kWh/m^3 $\times$ 0.94 kg CO_2e/kWh = 273 g CO_2e/m^3. Considering the indirect GHG emissions resulting from chemical usage and transport, the total

Table 1.4 Main emission factors for direct and indirect GHG emissions in WWTPs.

Item	EF	Remark	Reference
Direct GHG emissions			
CH_4	10.4% of produced CH_4 in AD	34 kg $CO_2e/kg\ CH_4$	Parravicini *et al.* (2016)
N_2O	0.001–2.59% of influent TN	298 kg $CO_2d/kg\ N_2O$ Rely on TN removal and fit the relationship of N_2O-N/TN influent (%) = −0.049 × TN removal (%) + 4.553	Parravicini *et al.* (2016); Yoshida *et al.* (2014)
Indirect GHG emissions			
Electricity supply	0.94 kg CO_2e/kWh	Coal-burning power plant	Lenzen (2008)
Chemical consumption	0.02 kg CO_2e/m^3	Including chemicals for phosphorus removal and sludge dewatering	Parravicini *et al.* (2016)
Transport	0.01 kg CO_2e/m^3		Parravicini *et al.* (2016)

indirect GHG emission could be up to 303 g CO_2e/m^3. Consequently, the total GHG emission in the Ulu Pandan WRP should stand at around $506 + 303 = 809$ g CO_2e/m^3, which is comparable with reported values in the literature (Molinos-Senante *et al.*, 2014). It should be pointed out that the GHG emissions from waste sludge disposal, sewer system and receiving bodies, etc., may also affect the carbon footprint analysis in WWTPs while they play minor roles in the carbon balance, and they are thus not included in the present analysis (Parravicini *et al.*, 2016).

Given average global municipal wastewater production of 360 km^3/year (Flörke *et al.*, 2013; Sato *et al.*, 2013), GHG emissions from WWTPs would stand at the level of 360 $km^3/year \times 809$ g $CO_2e/m^3 = 0.29$ G ton CO_2e/year. As global climate change is becoming a pressing issue (Burke *et al.*, 2018; Figueres *et al.*, 2018; Warren *et al.*, 2018), the emission of such a large amount of CO_2e will definitely pose a serious concern for the environmental sustainability of current wastewater treatment practices.

As illustrated in Figure 1.8, electricity supply to WWTP causes the most GHG emissions in WWTPs, accounting for 34% of emissions. Obviously, by improving the energy self-sufficiency of WWTPs the indirect GHG emissions

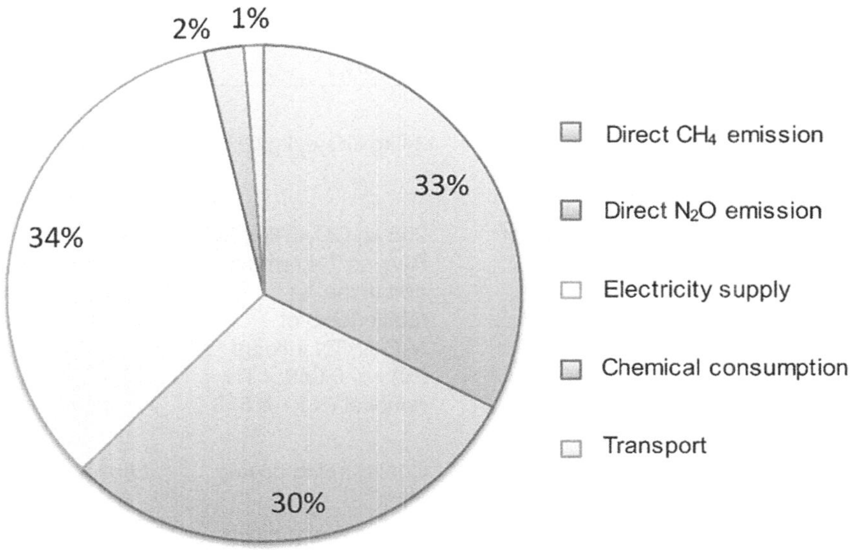

Figure 1.8 Analysis of carbon footprint in Ulu Pandan WRP.

can be significantly reduced or eventually eliminated, while direct CH_4 emission needs to be properly managed through improving CH_4 utilization efficiency for energy recovery. On the other hand, N_2O emission indeed plays an important role in the WWTPs carbon footprint, and advanced nitrogen removal technologies should be developed to improve TN removal efficiency with reduced N_2O production. Consequently, wastewater treatment, energy consumption and GHG emission in WWTPs should be viewed and addressed in a more holistic manner.

1.6 THE EMERGING SITUATION OF WASTE SLUDGE

1.6.1 Generation of waste sludge

With rapid global population growth and urbanization, the annual amount of global municipal wastewater production has been increasing. As such, a huge amount of waste sludge is generated as a byproduct of wastewater treatment and is becoming the most challenging problem for biological wastewater treatment (Drechsel *et al.*, 2015).

The waste sludge generated in WWTPs mainly originates from primary and secondary settling, where the primary sludge mainly consists of suspended solids and the secondary sludge refers to excess biomass produced during biological treatment of wastewater. During biological wastewater treatment, COD is partially converted to biomass via microbial assimilation, which is determined by sludge growth yield. In the widely used CAS process, the typical sludge growth yield had been reported to be in the range of 0.3–0.5 g dry biomass/g removed COD, indicating that 0.3–0.5 kg dry biomass would be generated by removing 1 kg COD. In fact, annual dry waste sludge generation already hit about 10 million in the European Union, while reaching 8 million tons in the USA in 2010 (Liu *et al.*, 2018). In China, waste sludge production continued to inrease at an average annual growth of 13% in the period 2007 to 2013, while dry waste sludge generation had already reached 6.25 million tons in 2013 (Yang *et al.*, 2015).

In WWTPs, the disposal of waste sludge is energy-intensive and costly, accounting for 20–30% of total energy consumption (Figure 1.4) and 25–65% of total operation costs (Liu and Tay, 2001). Moreover, heavy metals and pathogenic microbe in WAS will cause serious concerns on potential environmental risks if not treated appropriately (Zhang *et al.*, 2017). Therefore, the development of strategies for efficient and economic disposal of WAS is urgently needed.

1.6.2 Current technologies for waste sludge treatment

Volume reduction, non-toxicity and energy/resource recovery are the directive principles for waste sludge management (Zhang *et al.*, 2017). Landfill, sludge dewatering/concentration and incineration are the commonly used techniques for waste sludge treatment. Recently, more and more effort has been devoted to developing innovative technologies, e.g. high-performance anaerobic digestion, aerobic fermentation, pyrolysis and gasification, to improve sludge treatment efficiency as well as enhance energy/resource recovery and volume reduction. The main waste sludge treatment technologies are summarized in Figure 1.9.

Dewatering and concentration are the primary technologies for sludge treatment in WWTPs, which can help to relieve the workload of the subsequent treatment facilities and equipment, while improving the treatment efficiency by reducing the moisture content and volume of waste sludge (Vesilind, 1994). After such processing, about 75–80% of volume reduction can be achievable. Landfill is another traditional option for waste sludge disposal, in which waste sludge is stabilized naturally (Zhu & Zhao, 2011). Due to its advantages of easy operation, low cost and large treatment capacity, landfill is still one of the most popular technologies for waste sludge management in many countries. For example, 77% of waste sludge is currently disposed via landfill in the European Union, while this figure is 64% in the USA (Drechsel *et al.*, 2015). Incineration is a thorough waste sludge disposal method which can realize concurrent sludge volume reduction, non-toxicity, and energy/resource recovery, and it has been considered one of the promising technologies for waste sludge disposal in most developed countries (Kelessidis & Stasinakis, 2012; Murakami *et al.*, 2009).

Anaerobic digestion has been considered as a feasible and economic approach for sludge treatment in large-scale WWTPs with the merits of efficient energy recovery and low environmental pollution. However, it should be realized that only 30–50% of waste sludge can be degraded towards the production of biogas due to the low

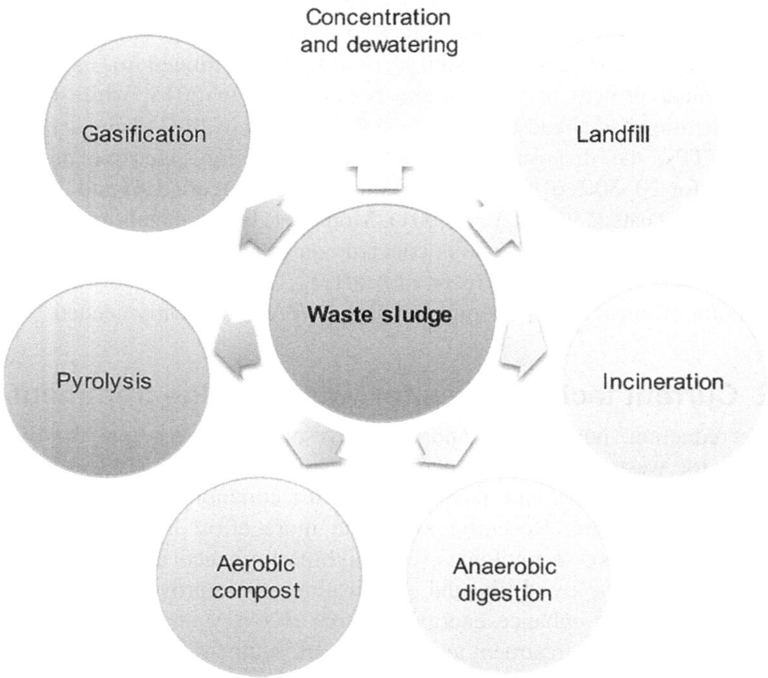

Figure 1.9 Current technologies for waste sludge treatment.

hydrolysis efficiency of waste sludge. To tackle this challenge, various pretreatment methods such as enzymatic, thermal, physical, chemical and mechanical pretreatments have been applied to accelerate the conversion of biosolids into soluble organics prior to anaerobic digestion, with the aim of improving the methane yield and sludge volume reduction (Carlsson *et al.*, 2015; Li *et al.*, 2017; Ma *et al.*, 2017; Ruffino *et al.*, 2015). Moreover, co-digestion of waste sludge with other organic wastes (e.g. food waste, fatty wastewater etc.) has also attracted more and more attention (Ma *et al.*, 2017; Yin *et al.*, 2016). Composting is a typical aerobic fermentation technology, which utilizes microbes to oxidize and decompose various organic matter into humus under controlled conditions, resulting in sludge stabilization (Yamada & Kawase, 2006). In general, compost product from waste sludge may be a safe biofertilizer for plant growth, with obvious social and environmental benefits. Pyrolysis refers to a very complex chemical reaction process of converting waste sludge into gaseous, liquid and solid products by heating under anaerobic or anoxic conditions (Kim & Parker, 2008). Pyrolysis products can be reused as energy and resources, e.g. pyrolysis

Table 1.5 Comparison of current technologies for waste sludge treatment.

Methods	Pros	Cons	References
Concentration and dewatering	Reduction of water content and volume of waste sludge; Lighten the subsequent workload; Improve waste sludge treatment efficiency.	Incomplete disposal.	Vesilind (1994)
Landfill	Easy operation and management; Low capital and operation costs; Large disposal capacity.	Potential secondary pollution; No energy and resource recovery; Large footprint.	Zhu and Zhao (2011)
Incineration	Easy operation; Complete disposal; Significant waste sludge volume reduction; High treatment efficiency; Energy and heat recovery.	High water content with low heating value; High capital and operation costs; Generation of harmful gas.	Hong *et al.* (2013)
Anaerobic digestion	High degree of automation; Low operation costs; Energy recovery.	High capital and maintenance costs; Subsequent processes for CH_4 to electricity; Long retention time (15–30 days).	Mao *et al.* (2015)
Aerobic compost	Easy operation and management; High degree of mechanization; High additional value of products; High resource utilization rate.	High costs for functional microorganisms; Leachate and nuisance odours; Long retention time (30–60 days); Large footprint.	Yamada and Kawase (2006)
Pyrolysis	Small footprint; Energy and resource recovery; Heavy metal stabilization; No Dioxin generated; Low GHG emissions.	Insignificant volume reduction; High costs for residue treatment; High investment costs; Suitable for large-scale plants.	Kim and Parker (2008)
Gasification	High energy efficiency; Low operation cost; Less harmful emissions; Heavy metal fixed ash residue with low leaching toxicity.	Waste sludge water content limited; High investment and operation costs; Production of organic pollutant compounds.	Gasafi *et al.* (2008)

solids residues can be used as adsorbents, while pyrolysis gases and oils can be collected as energy sources (Kim & Parker, 2008). In recent years, co-pyrolysis of waste sludge and other organic waste, e.g. household waste, lower-grade coals and biomass have been extensively explored. Sludge gasification involves a series of thermochemical reactions under specific temperature and pressure, using air, oxygen or water vapour as the gasification agent to convert organic matter in waste sludge into flammable gas (e.g. H_2, CH_4, CO etc.), tar and ash residues (He *et al.*, 2014). This technology offers the feasibility of direct use of fuel gas products from sludge gasification for energy recovery with high efficiency and lower leaching toxicity.

As discussed above, though many various technologies are currently being developed for waste sludge disposal, each of them has intrinsic shortcomings, as summarized in Table 1.5. With increasingly stringent legal regulations, there are serious concerns about rising energy consumption, operation costs and public sensitivity to current waste sludge disposal. However, it should be realized that reducing and minimizing sludge generation during wastewater treatment rather than post-handling of waste sludge should be more promising to solve this challenging issue at its roots.

REFERENCES

Ahn J. H., Kim S., Park H., Rahm B., Pagilla K. and Chandran K. (2010). N_2O emissions from activated sludge processes, 2008–2009: results of a national monitoring survey in the United States. *Environmental Science & Technology*, **44**(12), 4505–4511.

Appels L., Baeyens J., Degrève J. and Dewil R. (2008). Principles and potential of the anaerobic digestion of waste-activated sludge. *Progress in Energy and Combustion Science*, **34**(6), 755–781.

AQUASTAT (2018). AQUASTAT: FAO Global Online Information System on Water and Agriculture. Wastewater section, Vol. 2018. Food and Agriculture Organization of the United Nations (FAO), Washington DC, USA.

Armstrong R. C., Wolfram C., de Jong K. P., Gross R., Lewis N. S., Boardman B., Ragauskas A. J., Ehrhardt-Martinez K., Crabtree G. and Ramana M. V. (2016). The frontiers of energy. *Nature Energy*, **1**, 15020.

Burke M., Davis W. M. and Diffenbaugh N. S. (2018). Large potential reduction in economic damages under UN mitigation targets. *Nature*, **557**(7706), 549.

Cao Y. S. (2011). Mass Flow and Energy Efficiency of Municipal Wastewater Treatment Plants. IWA Publishing, London, UK.

Carlsson M., Holmstrom D., Bohn I., Bisaillon M., Morgan-Sagastume F. and Lagerkvist A. (2015). Impact of physical pre-treatment of source-sorted organic fraction of municipal solid waste on greenhouse-gas emissions and the economy in a Swedish anaerobic digestion system. *Waste Management*, **38**, 117–125.

China Ministry of Environmental Protection (China MEP) (2017). National Report on Environmental Quality of 2015. China MEP, China.

Crawford G., Johnson T., Johnson B., Krause T. and Wilner H. (2011). CHEApet Users Manual. Water Environment Research Foundation, Virginia, USA.

Daelman M. R. J., van Voorthuizen E. M., van Dongen U., Volcke E. I. P. and van Loosdrecht M. C. M. (2012). Methane emission during municipal wastewater treatment. *Water Research*, **46**(11), 3657–3670.

Drechsel P., Qadir M. and Wichelns D. (2015). Wastewater: Economic Asset in an Urbanizing World. Springer, Dordrecht.

Dudley B. (2018). BP statistical review of world energy 2017. London, UK.

Eggleston S., Buendia L., Miwa K., Ngara T. and Tanabe K. (2006). 2006 IPCC Guidelines for National Greenhouse Gas Inventories. Institute for Global Environmental Strategies Hayama, Japan.

Electric Power Research Institute (EPRI) (2002). Water & Sustainability (Volume 4): U.S. Electricity Consumption for Water Supply & Treatment – The Next Half Century Technical Report.

Enerdata (2018). Global Energy Statistical Yearbook 2018. See: https://yearbook.enerdata.net/ (accessed September 2019).

Figueres C., Le Quere C., Mahindra A., Baete O., Whiteman G., Peters G. and Guan D. (2018). Emissions are still rising: ramp up the cuts. *Nature*, **564**(7734), 27–30.

Flörke M., Kynast E., Bärlund I., Eisner S., Wimmer F. and Alcamo J. (2013). Domestic and industrial water uses of the past 60 years as a mirror of socio-economic development: A global simulation study. *Global Environmental Change*, **23**(1), 144–156.

Gans N., Mobini S. and Zhang X. (2007). Mass and energy balances at the Gaobeidian wastewater treatment plant in Beijing, China, Masters Thesis.

Gasafi E., Reinecke M.-Y., Kruse A. and Schebek L. (2008). Economic analysis of sewage sludge gasification in supercritical water for hydrogen production. *Biomass & Bioenergy*, **32**(12), 1085–1096.

Gellings C. W. and Parmenter K. E. (2016). Energy efficiency in fertilizer production and use. In: Efficient Use and Conservation of Energy, C. W. Gellings (ed.). EOLSS Publishing Co. Ltd, Oxford, UK.

Goldstein R. and Smith W. (2002). *Water & Sustainability (volume 4): US Electricity Consumption for Water Supply & Treatment-the Next Half Century*. Electric Power Research Institute.

Gu J., Yang Q. and Liu Y. (2018). Mainstream anammox in a novel A-2B process for energy-efficient municipal wastewater treatment with minimized sludge production. *Water Research*, **138**, 1–6.

Hao X., Liu R. and Huang X. (2015). Evaluation of the potential for operating carbon neutral WWTPs in China. *Water Research*, **87**, 424–431.

He C., Chen C.-L., Giannis A., Yang Y. and Wang J.-Y. (2014). Hydrothermal gasification of sewage sludge and model compounds for renewable hydrogen production: A review. *Renewable & Sustainable Energy Reviews*, **39**, 1127–1142.

Hellinga C., Schellen A., Mulder J. W., van Loosdrecht M. V. and Heijnen J. (1998). The SHARON process: an innovative method for nitrogen removal from ammonium-rich waste water. *Water Science and Technology*, **37**(9), 135–142.

Henze M., van Loosdrecht M. C., Ekama G. A. and Brdjanovic D. (2008). Biological Wastewater Treatment. IWA publishing.

Hong J., Xu C., Hong J., Tan X. and Chen W. (2013). Life cycle assessment of sewage sludge co-incineration in a coal-based power station. *Waste Management*, **33**(9), 1843–1852.

Jonasson M. and Ulf Jeppsson I. (2007). Energy Benchmark for Wastewater Treatment Processes: a comparison beetween Sweden and Austria, MS Thesis, 2007, Dept of Industrial Electrical Engineering and Automation, Lund University, Sweden.

Kartal B., Kuenen J. V. and Van Loosdrecht M. (2010). Sewage treatment with anammox. *Science*, **328**(5979), 702–703.

Keating D. (2018). EU carbon trading is working again – and power prices are rising. https://www.forbes.com/sites/davekeating/2018/08/31/eu-carbon-trading-is-working-again-and-power-prices-are-rising/#37b711041435 (accessed 31 August 2018).

Kelessidis A. and Stasinakis A. S. (2012). Comparative study of the methods used for treatment and final disposal of sewage sludge in European countries. *Waste Management*, **32**(6), 1186–1195.

Kim Y. and Parker W. (2008). A technical and economic evaluation of the pyrolysis of sewage sludge for the production of bio-oil. *Bioresource Technology*, **99**(5), 1409–1416.

Lee M., Keller A. A., Chiang P.-C., Den W., Wang H., Hou C.-H., Wu J., Wang X. and Yan J. (2017). Water-energy nexus for urban water systems: A comparative review on energy intensity and environmental impacts in relation to global water risks. *Applied Energy*, **205**, 589–601.

Lenzen M. (2008). Life cycle energy and greenhouse gas emissions of nuclear energy: A review. *Energy Conversion and Management*, **49**(8), 2178–2199.

Li N., Liu H., Xue Y., Wang H. and Dai X. (2017). Partition and fate analysis of fluoroquinolones in sewage sludge during anaerobic digestion with thermal hydrolysis pretreatment. *Science of the Total Environment*, **581**, 715–721.

Liu Y. J., Gu J. and Liu Y. (2018). Energy self-sufficient biological municipal wastewater reclamation: present status, challenges and solutions forward. *Bioresource Technology*, **269**, 513–519.

Liu Y. and Tay J.-H. (2001). Strategy for minimization of excess sludge production from the activated sludge process. *Biotechnology Advances*, **19**(2), 97–107.

Logan B. E. (2008). Microbial Fuel Cells. John Wiley & Sons.

Ma Y., Yin Y. and Liu Y. (2017). New insights into co-digestion of activated sludge and food waste: Biogas versus biofertilizer. *Bioresource Technology*, **241**, 448–453.

Mao C., Feng Y., Wang X. and Ren G. (2015). Review on research achievements of biogas from anaerobic digestion. *Renewable & Sustainable Energy Reviews*, **45**, 540–555.

Massara T. M., Malamis S., Guisasola A., Baeza J. A., Noutsopoulos C. and Katsou E. (2017). A review on nitrous oxide (N_2O) emissions during biological nutrient removal from municipal wastewater and sludge reject water. *Science of The Total Environment*, **596–597**, 106–123.

McCarty P. L., Bae J. and Kim J. (2011). Domestic wastewater treatment as a net energy producer–can this be achieved? *Environmental Science & Technology*, **45**, 7100–7106.

Mihelcic J. R., Zimmerman J. B. and Auer M. T. (2014). Environmental Engineering: Fundamentals, Sustainability, Design. Wiley, Hoboken.

Mizuta K. and Shimada M. (2010). Benchmarking energy consumption in municipal wastewater treatment plants in Japan. *Water Science and Technology*, **62**(10), 2256.

Molinos-Senante M., Hernandez-Sancho F., Mocholi-Arce M. and Sala-Garrido R. (2014). Economic and environmental performance of wastewater treatment plants: Potential reductions in greenhouse gases emissions. *Resource and Energy Economics*, **38**, 125–140.

Monteith H. D., Sahely H. R., MacLean H. L. and Bagley D. M. (2005). A rational procedure for estimation of greenhouse-gas emissions from municipal wastewater treatment plants. *Water Environment Research*, **77**(4), 390–403.

Monteith H., Kalogo Y. and Louzeiro N. (2007). Achieving stringent effluent limits takes a lot of energy! *Proceedings of the Water Environment Federation*, **2007**(14), 4343–4356.

Murakami T., Suzuki Y., Nagasawa H., Yamamoto T., Koseki T., Hirose H. and Okamoto S. (2009). Combustion characteristics of sewage sludge in an incineration plant for energy recovery. *Fuel Processing Technology*, **90**(6), 778–783.

Myhre G., Shindell D., Bréon F. M., Collins W., Fuglestvedt J., Huang J., Koch D., Lamarque J. F., Lee D. and Mendoza B. (2013). Anthropogenic and natural radiative forcing. *Climate Change*, **423**, 658–740.

Nouri J., Jafarinia M., Naddafi K., Nabizadeh R., Mahvi A. and Nouri N. (2006). Energy recovery from wastewater treatment plant. *Pakistan Journal of Biological Sciences*, **9** (1), 3–6.

Pabi S., Amarnath A., Goldstein R. and Reekie L. (2013). Electricity Use and Management in the Municipal Water Supply and Wastewater Industries. Water Research Foundation, Denver, CO, USA.

Parravicini V., Svardal K. and Krampe J. (2016). Greenhouse gas emissions from wastewater treatment plants. In: European Geosciences Union General Assembly 2016, M. Ask, V. Bruckman, C. Juhlin, T. Kempka and M. Kuhn (eds.), Elsevier Science Bv., Amsterdam, Vol. **97**, pp. 246–253.

Qi C. and Chang N.-B. (2013). Integrated carbon footprint and cost evaluation of a drinking water infrastructure system for screening expansion alternatives. *Journal of Cleaner Production*, **60**, 170–181.

Ruffino B., Campo G., Genon G., Lorenzi E., Novarino D., Scibilia G. and Zanetti M. (2015). Improvement of anaerobic digestion of sewage sludge in a wastewater treatment plant by means of mechanical and thermal pre-treatments: Performance, energy and economical assessment. *Bioresource Technology*, **175**, 298–308.

Sabba F., Terada A., Wells G., Smets B. F. and Nerenberg R. (2018). Nitrous oxide emissions from biofilm processes for wastewater treatment. *Applied Microbiology and Biotechnology*, **102**(22), 9815–9829.

Sato T., Qadir M., Yamamoto S., Endo T. and Zahoor A. (2013). Global, regional, and country level need for data on wastewater generation, treatment, and use. *Agricultural Water Management*, **130**, 1–13.

Sawin J. L., Rutovitz J. and Sverrisson F. (2018). Renewables 2018–Global status report. See: ren21.net/gsr-2018/ (accessed September 2019).

Scherson Y. D., Wells G. F., Woo S.-G., Lee J., Park J., Cantwell B. J. and Criddle C. S. (2013). Nitrogen removal with energy recovery through N_2O decomposition. *Energy & Environmental Science*, **6**(1), 241–248.

Schreiber F., Wunderlin P., Udert K. M. and Wells G. F. (2012). Nitric oxide and nitrous oxide turnover in natural and engineered microbial communities: biological pathways, chemical reactions, and novel technologies. *Frontiers in Microbiology*, **3**, 372.

Shapleigh J. P. (2013). Denitrifying prokaryotes. In: The Prokaryotes, E. Rosenberg, E. F. DeLong, S. Lory, E. Stackebrandt and F. Thompson (eds), Springer, Berlin, Heidelberg, pp. 405–425.

Tchobanoglous G., Burton F. L. and Stensel H. D. (2003). Wastewater Engineering Treatment and Reuse. McGraw-Hill Higher Education, Boston, USA.

UNESCO (2012). World Water Development Report 4—Managing Water Under Uncertainty and Risk. UNESCO, New York, USA.

United States Environmental Protection Agency (US EPA) (2006). Wastewater Management Fact Sheet, Energy Conservation. EPA 832-F-06-024. EPA, Washington DC, USA.

United States Environmental Protection Agency (US EPA) (2007). Opportunities for and Benefits of Combined Heat and Power at Wastewater Treatment Facilities. EPA-430-R-07-003, 6. EPA, Washington DC, USA.

Van Dongen U., Jetten M. S. and Van Loosdrecht M. C. (2001). The SHARON®-Anammox® process for treatment of ammonium rich wastewater. *Water Science and Technology*, **44**(1), 153.

Verstraete W., Van de Caveye P. and Diamantis V. (2009). Maximum use of resources present in domestic "used water". *Bioresource technology*, **100**(23), 5537–5545.

Vesilind P. A. (1994). The role of water in sludge dewatering. *Water Environment Research*, **66**(1), 4–11.

Wan J., Gu J., Zhao Q. and Liu Y. (2016). COD capture: a feasible option towards energy self-sufficient domestic wastewater treatment. *Scientific Reports*, **6**, 25054.

Warren R., Price J., Graham E., Forstenhaeusler N. and VanDerWal J. (2018). The projected effect on insects, vertebrates, and plants of limiting global warming to 1.5°C rather than 2°C. *Science*, **360**(6390), 791–795.

Wunderlin P., Mohn J., Joss A., Emmenegger L. and Siegrist H. (2012). Mechanisms of N_2O production in biological wastewater treatment under nitrifying and denitrifying conditions. *Water Research*, **46**(4), 1027–1037.

Yamada Y. and Kawase Y. (2006). Aerobic composting of waste activated sludge: Kinetic analysis for microbiological reaction and oxygen consumption. *Waste Management*, **26**(1), 49–61.

Yang G., Zhang G. and Wang H. (2015). Current state of sludge production, management, treatment and disposal in China. *Water Research*, **78**, 60–73.

Yin Y., Liu Y.-J., Meng S.-J., Kiran E. U. and Liu Y. (2016). Enzymatic pretreatment of activated sludge, food waste and their mixture for enhanced bioenergy recovery and waste volume reduction via anaerobic digestion. *Applied Energy*, **179**, 1131–1137.

Yoshida H., Monster J. and Scheutz C. (2014). Plant-integrated measurement of greenhouse gas emissions from a municipal wastewater treatment plant. *Water Research*, **61**, 108–118.

Zhang Q., Yang W., Ngo H., Guo W., Jin P., Dzakpasu M., Yang S., Wang Q., Wang X. and Ao D. (2016). Current status of urban wastewater treatment plants in China. *Environment International*, **92**, 11–22.

Zhang Q., Hu J., Lee D.-J., Chang Y. and Lee Y.-J. (2017). Sludge treatment: Current research trends. *Bioresource Technology*, **243**, 1159–1172.

Zhou Y., Zhang D. Q., Le M. T., Puah A. N. and Ng W. J. (2013). Energy utilization in sewage treatment–a review with comparisons. *Journal of Water and Climate Change*, **4**(1), 1–10.

Zhu Y. and Zhao Y. (2011). Stabilization process within a sewage sludge landfill determined through both particle size distribution and content of humic substances as well as by FT-IR analysis. *Waste Management & Research*, **29**(4), 379–385.

Chapter 2

Approaches to energy and resource recovery from municipal wastewater

Meng Zhang

2.1 INTRODUCTION

The activated sludge process is regarded by many as a remarkable engineering development of the 20th Century which has made a great contribution to wastewater treatment. But after more than 100 years of successful application, the conventional activated sludge (CAS) process has recently been receiving an increasing number of critiques due to the high energy consumption and excessive sludge generation associated with it. Increasing efforts have been devoted to energy recovery in the form of biogas produced from waste sludge through anaerobic digestion (AD). However, the recoverable electrical energy generated from the AD of waste sludge can only offset about 50–60% of the total input energy of wastewater treatment plants (WWTPs) with CAS as the core process, while energy efficiency could be increased further to about 75% with upgraded AD equipped with pre-treatment of thickening sludge and a combined heat and power (CHP) process (Cao, 2011). Moreover, process upgrading and retrofitting are urgently needed in more and more WWTPs around the world to meet the tightened effluent discharge standards applied in recent years which inevitably lead to a rising trend of in-plant energy consumption. For example, it has been reported that the effluent discharge standards in China have been gradually migrated to quasi Class IV for surface water bodies with very low discharge limits. Consequently, it appears almost impossible or highly challenging to

Author(s): Yu Liu, Jun Gu and Meng Zhang
doi: 10.2166/9781789060089_0029

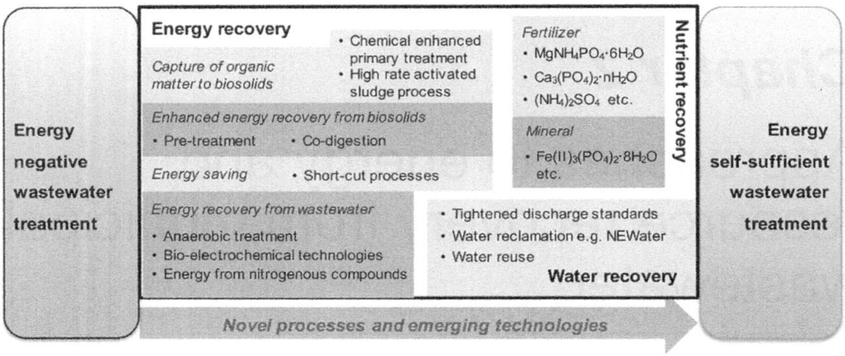

Figure 2.1 Roadmap towards energy self-sufficient municipal wastewater treatment.

realize energy self-sufficient municipal wastewater treatment through further process optimization if the CAS process continues to serve as the core biological unit.

Facing such challenging situations, instead of continuing with the current practice it is necessary to explore novel process configurations and emerging technologies. The solutions moving forward should rely on ways to enhance energy recovery from biosolids and wastewater, while minimizing energy consumption. This chapter thus presents and elucidates a roadmap of emerging technologies with potentials for energy recovery and for energy saving together with possible resource recovery, as a way to move towards the ultimate goal of energy self-sufficient municipal wastewater treatment (Figure 2.1).

2.2 TECHNOLOGIES FOR ENERGY RECOVERY FROM BIOSOLIDS

Organic matter in wastewater, usually expressed in terms of COD, has been widely explored as an important source of recoverable energy in wastewater treatment processes. The recoverable energy embedded in organic matter accounts for about 86% of the total potential energy in municipal wastewater (Table 1.1). However, about 40–50% of soluble COD in wastewater is oxidized to carbon dioxide in current biological processes for municipal wastewater treatment, which is designed and operated on the basis of biological oxidation. As such, this amount of chemical energy in municipal wastewater is eventually lost. Obviously, direct capture of COD prior to bio-oxidation may offer a feasible engineering solution for maximizing the energy recovery. This may likely lead to a paradigm shift for current WWTPs with benefits including: (i) improved energy recovery; (ii) reduced energy consumption associated with aeration; and (iii) minimized waste sludge production (Wan *et al.*, 2016). Currently, COD is usually channeled into biosolids during biological treatment of wastewater and then directed to AD for methane production in the sidestream sludge line.

2.2.1 Technologies for the capture of organic matter

2.2.1.1 The chemically enhanced primary treatment process

It has been widely accepted that the primary settling in the CAS process generally offers an efficient means of quickly removing settleable solids, while most of the suspended particulates, colloidal and soluble organic matters enter the subsequent biological process (Metcalf & Eddy, 2014). Unlike the conventional primary clarifier, chemically enhanced primary treatment (CEPT) has been developed to improve the removal of inorganic and organic solids by the addition of chemical coagulants to wastewater (e.g. iron salts, aluminium salts and organic polyelectrolytes). Through coagulation and flocculation, small suspended solids can be agglomerated into large flocs, leading to enhanced removal of suspended solids from wastewater (Chagnon & Harleman, 2002).

The CEPT process basically includes a small rapid mixing tank for coagulant dosing, a slow mixing flocculation tank and a clarification tank, which can be integrated with the existing primary settler. The dose of chemicals is the main factor affecting the performance of COD capture by CEPT. Typically, iron and aluminium salts, or organic polyelectrolytes, are used at the doses of 100 mg/L, 80 mg/L and 0.5–4 mg/L, respectively (Diamantis *et al.*, 2011). It should be noted that the actual dosage of a coagulant is dependent on the wastewater composition. It has been found that 70–80% of COD, 90% of suspended solids and 90% of phosphorus together with some heavy metals can be removed through CEPT (Diamantis *et al.*, 2011), while up to 50% of soluble COD could also be captured via adsorption of bioflocs. Moreover, integrated CEPT with dissolved air flotation could lead to more efficient harvesting of highly concentrated sludge, favouring subsequent AD for improved biogas production.

The COD recovered from the primary sludge could be increased from 26% in the CAS process to 39% of the total incoming COD in the process coupled with CEPT (Wan *et al.*, 2016). Theoretically, for municipal wastewater with a COD concentration of 500 mg/L, about 195 grams of methane-COD could be recoverable from one cubic meter of wastewater. Considering the chemical energy of methane-COD being about 13.91 kJ/g (Heidrich *et al.*, 2010), the unit recoverable energy from COD captured by CEPT can be estimated as 195 $g/m^3 \times 13.91\ kJ/g = 2{,}712\ kJ/m^3$. At an electricity conversion efficiency of 35%, the total electrical energy of about 949 kJ/m^3 (i.e. 0.26 kWh/m^3) would be ultimately recovered via AD of the primary sludge harvested by CEPT.

Although CEPT was developed decades ago, it has not yet gained the popularity initially expected of it in municipal wastewater treatment due to the use of chemical coagulants, high operation cost, increased volume of total solids (TS), etc. Evidence suggested that the operating and maintenance costs of CEPT were higher than those of conventional primary treatment (Chagnon & Harleman, 2002), e.g. the operating costs of CEPT increased by about €0.06–0.10/m^3 wastewater treated, equivalent to €0.24–0.40/kg COD due to the use of chemicals (Diamantis *et al.*, 2013). Moreover, the higher dosage of chemicals, especially at concentrations higher than 100 mg/L for iron or alum, might compromise the biogas productivity in the subsequent AD process. On the other hand, the addition of coagulants during CEPT would also lead to an increased amount of TS by a factor of 1.5–2.0 against that produced in the conventional primary treatment process (Huang & Li, 2000). As such, proper disposal of such solids containing trivalent ions (e.g. Fe^{3+} and Al^{3+}) is obviously needed, and this in turn may cause additional costs associated with post-solid management (Xu *et al.*, 2009).

2.2.1.2 The high rate activated sludge process

The high rate activated sludge (HRAS) process offers a technical option for capturing particulate, colloidal and soluble organic matters from municipal wastewater via biosoprtion and bioaccumulation. As a result, a large amount of sludge is expected to be produced and can be further subjected to AD for biogas generation (Modin *et al.*, 2016). In fact, the working principle of HRAS is built on the philosophy of maximizing capture of particulate, colloidal and soluble COD, with minimized hydrolysis and mineralization of wastewater COD. For this purpose, a shorter hydraulic retention time (HRT) and sludge retention time (SRT) have commonly been adopted under a limiting dissolved oxygen (DO) condition (Modin *et al.*, 2016; Rahman *et al.*, 2014). The biosludge from the HRAS process, which is known as adsorptive sludge or young sludge, is more digestible than the waste activated sludge (WAS) produced from the CAS process. Consequently, higher energy recovery from such sludge via AD is to be expected.

So far, HRAS processes have been designed in two configurations: adsorption/biooxidation and contact stabilization (Figure 2.2). The adsorption/biooxidation process consists of a high organic loaded activated sludge unit and a low organic loaded activated sludge unit, with separated clarifiers for each stage (Boehnke *et al.*, 1998; Sancho *et al.*, 2019). In general, the adsorption stage is fed with a very high organic loading rate (e.g. with a food to microorganism (F/M) ratio of 2–10 g BOD/(g $VSS^{-1} \cdot d^{-1}$),(where VSS is volatile suspended solids) and is operated at a short HRT of 15–30 min and an extremely short SRT of 3–12 hours. Under such dynamic conditions, organic matter is removed primarily through adsorption by activated sludge flocs instead of microbial assimilation and mineralization. The biooxidation stage is mainly designed for biological oxidation and nutrient removal, and is usually operated with an F/M ratio lower than 0.1 g BOD/g $VSS^{-1} \cdot d^{-1}$ and a long SRT of 8–20 days (Modin *et al.*, 2016; Van Winckel, 2014).

By contrast, in the contact stabilization process (Figure 2.2a), wastewater is mixed with activated sludge in the contact tank operated at a short HRT of 15–60 min under limited DO concentration (Liu *et al.*, 2009), where COD is mainly removed by biological adsorption and storage under the dynamic conditions in the contact tank. After clarification, the concentrated sludge is

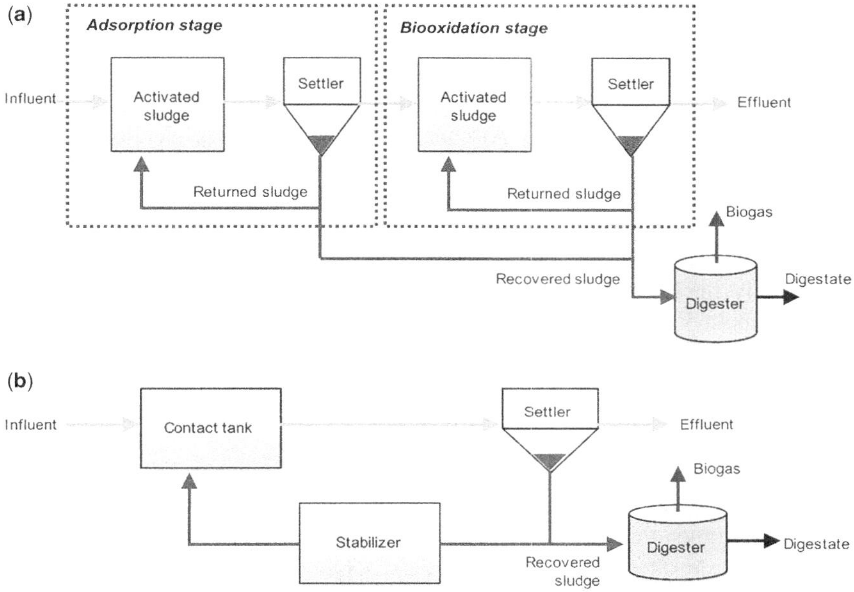

Figure 2.2 Schematics of typical HRAS processes. (a) adsorption/biooxidation process and (b) contact stabilization process. Adapted from Sancho *et al.* (2019), with permission.

returned to an aerated stabilizer where adsorbed COD is biologically oxidized. In this configuration, COD is mainly removed in the contact tank, while the stabilization tank helps to improve biosorption of organic matter by using recycled sludge (Rahman *et al.*, 2017). It has been reported that sludge subjected to a feast–famine cycle could possess a higher COD adsorption capacity and storage in the stabilization tank (Dolejs *et al.*, 2016). Based on these features of contact-stabilization, a high-rate contact stabilization process has been developed for enhanced pre-concentration of wastewater COD (Meerburg *et al.*, 2015; Zhao *et al.*, 2000).

Extensive efforts have been devoted to exploring the operational conditions favourable for HRAS to achieve a more efficient COD capture. In this regard, HRT, SRT, DO and the F/M ratio have been identified as key parameters for both the adsorption/biooxidation and contact-stabilization processes, with an optimal range of $\leq$0.5 h for HRT, 0.1–1.0 d for SRT, 0.5–1.0 mg/L for DO concentration and 2–10 g BOD/(g VSS · d) for the F/M ratio (Sancho *et al.*, 2019). Short HRT and SRT can facilitate biosorption of COD by activated sludge, while helping to control microbial assimilation and carbon mineralization. In addition, these conditions may also help to reduce operating costs and the process footprint. In general, HRAS is operated under limited DO concentration in order to minimize the biological oxidation of COD to carbon dioxide

(Elliott, 2016). The F/M ratio has been identified as a critical parameter for the design and operation of an HRAS process, i.e. a higher F/M ratio may promote more sludge production, which is beneficial for the subsequent AD for energy recovery (Henze *et al.*, 2008). In the operation of HRAS processes, it is critical to balance the sludge directed to AD for enhanced biogas production and that returned for COD capture. As illustrated in Figure 2.2, there is a practical tradeoff between returned sludge for COD capture and harvested sludge for AD, i.e. more sludge for AD, less sludge for COD capture. In fact, such a situation poses a concern about the long-term process stability.

Generally, the HRAS process could capture about 55–65% of influent total COD including about 30% of soluble COD, while this figure could be further increased to 80% under optimal conditions (Sancho *et al.*, 2019; Wett *et al.*, 2007). Meanwhile, 80–95% of suspended solids (SS) and a small part of nitrogen could also be simultaneously removed in the HRAS process. If 65% of influent COD was primarily captured by sludge, and 65% of that was converted to methane via AD (Miron *et al.*, 2000; Parkin & Owen, 1986), thus about 42% of influent total COD could be ultimately harvested as methane in the HRAS process, which is about 62% higher than that achieved in the CAS process (Wan *et al.*, 2016). Municipal wastewater usually has a typical COD concentration of about 500 mg/L (Mihelcic *et al.*, 2014). Considering the recoverable energy from methane-COD being 13.91 kJ/g, the unit recoverable energy from COD captured by the HRAS process could be estimated as $500\ g/m^3 \times 42\% \times 13.91\ kJ/g = 2{,}938\ kJ/m^3$. Furthermore, given an electricity conversion efficiency of 35%, the electrical energy recoverable from the AD of sludge harvested from the HRAS process could be at a level of about $1{,}028\ kJ/m^3$, i.e. $0.29\ kWh/m^3$, which is about 62% of the energy used to derive the CAS process.

2.2.2 Enhanced energy recovery from biosolids

AD of biosolids is a well-developed technology for energy recovery in the form of biogas. However, the recoverable electrical energy from the AD of biosolids could only offset about 50–60% of the total energy consumption of WWTPs. As such, different pre-treatment methods and co-digestion with other organic substances have been extensively explored with the aim of improving the overall biogas production.

2.2.2.1 Pre-treatment of biosolids

AD is a highly complex biological process with a series of sequential reactions carried out by different functional microbial species, i.e. hydrolysis, acidogenesis, acetogenesis and methanogenesis, among which hydrolysis has been identified to be the rate-limiting step (Henze *et al.*, 2008). For this reason, the central point of pre-treatment is to improve the hydrolysis of complex biosolids in order to break-up the bottleneck of AD. As depicted in Figure 2.3, the pre-treatment of

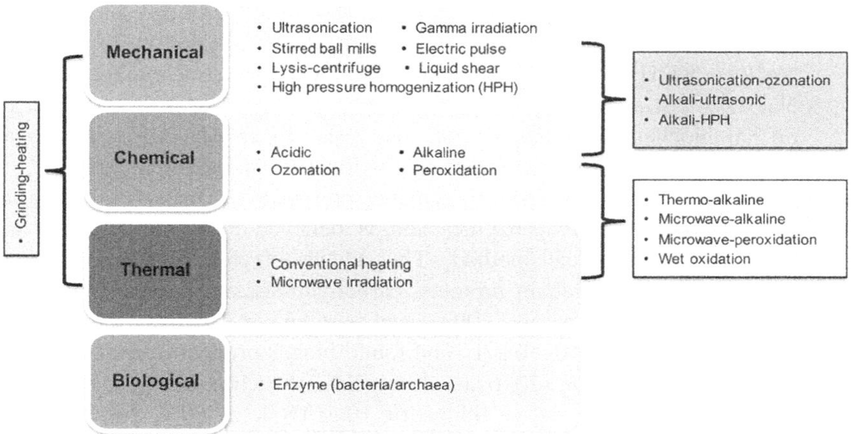

Figure 2.3 Pre-treatment methods of biosolids towards enhanced AD. Information from Gonzalez *et al.* (2018) and González *et al.* (2018).

biosolids can be performed through mechanical, chemical, biological or thermal methods, or by their combinations:

- **Thermal pretreatment:** During thermal pretreatment, the concentrated sludge is subjected to a high temperature of 150–230°C in a hydrolysis reactor. For waste sludge pretreated at 175°C for 60 minutes, it was found that gas production was increased from 108 to 216 mL/g COD (Li & Noike, 1992). To lower the temperature required, Maspolim *et al.* (2016) proposed a combined thermo-alkaline (55°C, pH 8.5) method, and found that 46.4 $\pm$ 4.2% of COD removal and 41.2 $\pm$ 4.0% of VSS reduction with a methane yield of 0.177 $\pm$ 0.017 L/g COD were achievable.
- **Mechanical pretreatment:** Ultrasound has been applied for sludge pretreatment at an energy input of 1,000–16,000 kJ/kg TS. After ultrasonic pretreatment, biogas production could be increased by 10%–45% in continuous or semi-continuous systems and by 24%–140% in batch systems (Carrère *et al.*, 2010). Lysis-centrifugation has also been applied to pretreat the thickened sludge from the dewatering centrifuges. This approach has been adopted in many WWTPs as a pretreatment for AD with a 15–26% increase in biogas production. In addition, liquid shear created by high pressure may also lead to strong mechanical disruption to cells and flocs. Engelhart *et al.* (2000) reported pretreatment of waste sludge at 300 bars, leading to increased CH_4 production from 129 to 206 mL/g VSS.
- **Chemical pretreatment:** Ozonation has been used to combine with AD as a pre-treatment step. After pretreatment of waste sludge with 0.15 g O_3/g TS, biogas production was increased from 150 to 367 mL/g influent COD

(Bougrier *et al.*, 2007). Alkali treatment is also used for sludge solubilization with good performance, and the effects of alkali salts follow the order of NaOH > KOH > $Mg(OH)_2$ and $Ca(OH)_2$. At an alkaline dosage of 0.3 g NaOH/g VS, methane generation was found to increase from 68.9 to 135 mL in biochemical methane potential tests (Li *et al.*, 2015). Hydrogen peroxide has also been explored as an oxidant for sludge pretreatment.

- **Enzymatic pretreatment:** Enzymatic pretreatment makes full use of the catalytic nature of enzymes and is believed to be an efficient and eco-friendly biological method. The added enzymes can improve the degradation of long-chain proteins, carbohydrates and lipids in biosolids. Barjenbruch and Kopplow (2003) used pure enzyme carbohydrase to treat thickened sludge of 40–50 g/L, and found biogas production was increased by 10–15%. Yin *et al.* (2016) used fungal mash rich in different hydrolytic enzymes to pretreat waste sludge of 10 g TS/L at 60°C for 24 h, and methane production was increased from 239.9 to 366.8 mL/g VS; similar results were also reported by Ma *et al.* (2018).

2.2.2.2 Co-digestion of biosolids

Co-digestion is a newly developed process with the concept of combining two or more substrates for simultaneous digestion in an anaerobic digester. Since anaerobic digesters in WWTPs are oversized (frequently by as much as 30%), this makes the addition of other biosolids together with waste sludge possible for anaerobic co-digestion (AcoD), resulting in improved biogas production and increased energy recovery (Lema & Martinez, 2017). In practice, waste sludge produced in WWTPs is the primary substrate for AcoD, but it has a relatively low C/N ratio and digestibility. Therefore, it has been proposed that waste sludge should be mixed with other biosolids rich in biodegradable organic matter (e.g. fats, oil and greases, food waste, algae, fruit and vegetable waste) for anaerobic co-digestion. According to the current technology readiness level (TRL) scale, AcoD has been assessed to be TRL 9, meaning that this technology is ready for large-scale commercialization (Edwards *et al.*, 2017). So far, AcoD has already been extensively applied as a viable option in many WWTPs worldwide.

Food waste is a kind of organic solid rich in carbohydrates, and typical biogas production from food waste is in the range of 0.7–0.9 Nm^3/kg VS, which is much higher than that from waste sludge generated in WWTPs (Koch *et al.*, 2016). Given such a situation, anaerobic co-digestion of waste sludge and food waste seems to be a feasible and viable approach for improving energy recovery towards energy self-sufficient wastewater reclamation. For example, after the addition of 1,000 m^3 of food waste into 9,000 m^3 of waste sludge the total biogas production was doubled (Koch *et al.*, 2016). Similarly, during the co-digestion of waste sludge with food waste pretreated with fungal mash, the methane generation was significantly increased from 360 mL CH_4/g VS for AD of waste

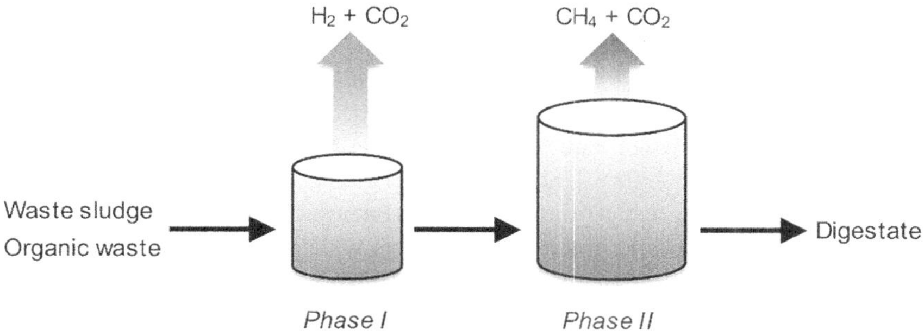

Figure 2.4 Schematic of a two-phase AcoD process. Adapted from Edwards *et al.* (2017), with permission.

sludge to 600–750 mL CH_4/g VS (Ma *et al.*, 2017; Yin *et al.*, 2016). Energy analysis further revealed that the unit recoverable energy reached 0.67–1.18 kWh/m^3 in AcoD, which was higher than the total energy consumed for wastewater reclamation (Ma *et al.*, 2017). These figures imply that current WWTPs may be transformed into energy producers instead of being consumers if co-digestion could be adopted.

A two-phase process has been proposed for AcoD as illustrated in Figure 2.4. In phase I, the reactor is operated with a short hydraulic retention time of less than 3 days and a high organic loading rate greater than 12 kg VS/($m^3 \cdot d$) to produce biogas rich in hydrogen, while the produced liquid rich in volatile fatty acids was then fed to the phase II reactor for biogas production, as usual (Edwards *et al.*, 2017). Gottardo *et al.* (2015) also reported a two-phase co-digestion system of the organic fraction of municipal solid waste and waste sludge, and found that the produced biogas constituted 58.6% methane, 7.5% hydrogen and 33.9% carbon dioxide. It should also be realized that co-digestion may affect the design and operation of several treatment facility stages, e.g. pretreatment, mixing, heating, dewatering etc. Definitely, transport and storage of co-substrates may impose additional capital investment and operating costs, while also causing environmental concerns (e.g. odour).

2.3 TECHNOLOGIES FOR ENERGY RECOVERY FROM WASTEWATER

A substantial amount of energy is embedded in wastewater. So far, anaerobic and bio-electrochemical technologies have been explored for direct or indirect energy recovery from organic matter in wastewater without generation of waste sludge. Nitrogenous compounds are also potential energy sources and are recoverable from municipal wastewater. Generally, energy in wastewater may be recovered in the forms of methane, ammonia, nitrous oxide, hydrazine, etc.

2.3.1 Anaerobic treatment of wastewater

Anaerobic treatment is a biodegradation process of organic matter through a series of biochemical reactions, including hydrolysis, acidogenesis, acetogenesis and methanogenesis which are carried out by different functional microbial species (Henze *et al.*, 2008). Unlike with CEPT and HRAS processes, influent COD is directly converted to biogas during anaerobic treatment of wastewater. Although various anaerobic processes have been widely applied for high-strength industrial wastewater treatment, increasing interest has recently also been dedicated to the use of the anaerobic process as a core unit for municipal wastewater treatment, known as mainstream anaerobic treatment, with the aims of concurrent energy recovery and sludge reduction (Liu *et al.*, 2018). Compared with the CAS process, anaerobic treatment of municipal wastewater has the advantages of energy generation, no aeration and less sludge production (Lema & Martinez, 2017). Theoretically, one gram of COD oxidized in the CAS process requires one gram of dissolved oxygen, equivalent to about 1 kWh of electrical energy consumed per kg of COD removed. Obviously, this energy associated with biological oxidation of COD can be totally saved in the anaerobic treatment of wastewater, while about 0.27–0.30 m^3 CH_4/kg COD removed could be generated during the anaerobic treatment of municipal wastewater. In general, the growth yield of anaerobic sludge in the mainstream anaerobic treatment process falls into the range of 0.03–0.20 g dry biomass/g COD removed, which is only about 10–33% of that in the CAS process (e.g. 0.3–0.5 g dry biomass/g COD removed).

So far, various anaerobic processes – e.g. the up-flow anaerobic sludge blanket (UASB), expanded granular sludge bed (EGSB), anaerobic fixed bed reactor (AFBR) and the anaerobic membrane bioreactor (AnMBR) processes – have been developed for mainstream municipal wastewater treatment. About 60–90% of COD removal was achieved in a full-scale UASB reactor (810 m^3) treating domestic sewage (Florencio *et al.*, 2001), while 81.5% of COD was removed in a pilot-scale EGSB reactor for municipal wastewater treatment at a short HRT of 4 hours, and the effluent COD was below 50 mg/L (Dai *et al.*, 2011). Gu *et al.* (2018) developed a lab-scale AFBR to treat municipal wastewater with 92% of COD removal.

The AnMBR process, as a combination of anaerobic treatment and membrane separation, has attracted growing interest for municipal wastewater treatment. Due to the effective retention of biomass by the membrane, SRT and HRT can be decoupled in an AnMBR with a typical HRT of several hours (Lema & Martinez, 2017). Similar to an aerobic membrane bioreactor (MBR), an AnMBR can be designed and operated in submerged or sidestream mode, as illustrated in Figure 2.5 (Robles Martínez, 2013). Generally, sidestream AnMBRs have been mostly applied to treat high-strength industrial wastewater, whilst submerged AnMBRs are more suitable for treating low-strength municipal wastewater (Judd

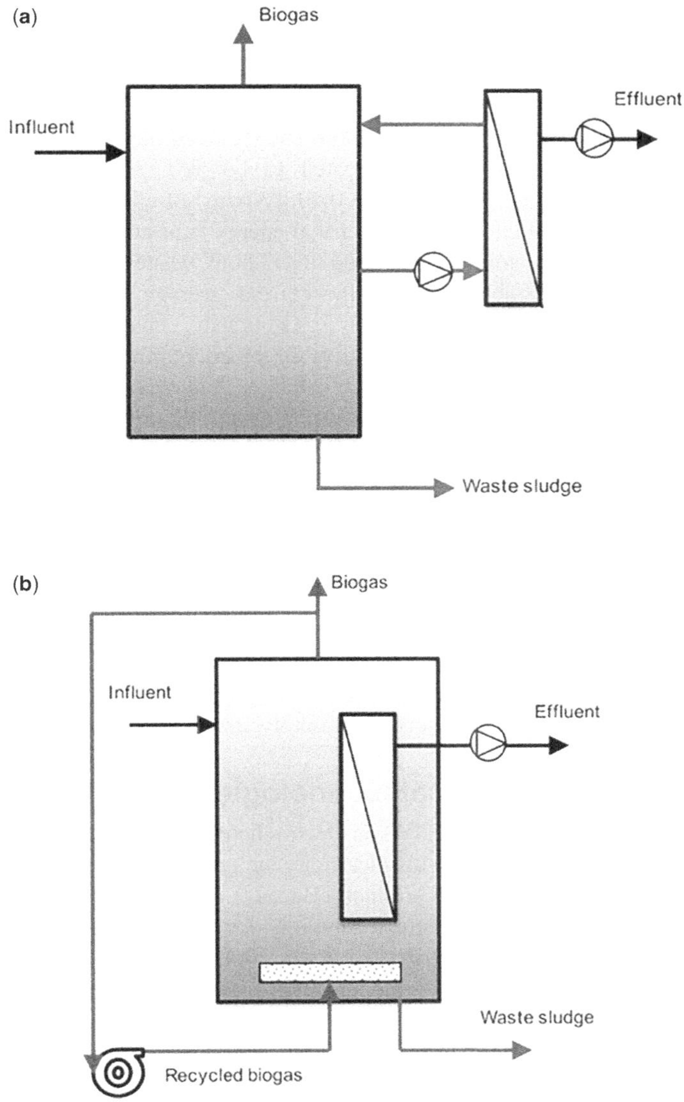

Figure 2.5 Schematic configurations of (a) a sidestream AnMBR and (b) a submerged AnMBR. Information taken from Robles Martínez (2013).

& Judd, 2010). Lema and Martinez (2017) reviewed the performances of 11 pilot-scale AnMBRs, including one sidestream and ten submerged configurations for municipal wastewater treatment. It turned out that 80–95% of COD could be removed at an HRT of 4.6–24.5 hours and with an SRT of 30–180 days.

Existing evidence suggests that over 80% of wastewater COD can be removed in the anaerobic processes described above. In these processes, the COD recovered from wastewater and biosludge accounted for 65% and 8% of the total COD, respectively (Wan *et al.*, 2016). For a typical municipal wastewater with COD concentration of 500 mg/L, the recoverable energy from the anaerobic treatment of municipal wastewater can be estimated to be $500\ g/m^3 \times (65\% + 8\%) \times 13.91\ kJ/g = 5{,}077\ kJ/m^3$, equivalent to $0.49\ kWh/m^3$ of electrical energy, if the conversion efficiency of chemical to electrical energy is assumed to be 35%. This implies that the energy jointly recovered from both wastewater and biosludge anaerobically would be able to offset the current average energy consumption in the CAS process (i.e. $0.47\ kWh/m^3$). Obviously, anaerobic treatment of wastewater offers a promising option for direct energy recovery without the generation of a large amount of excess biosludge. However, there are still some barriers that limit the widespread application of anaerobic processes for municipal wastewater treatment, of which dissolved methane is known as a big hurdle for anaerobic treatment of low-strength municipal wastewater. It is estimated that about half of produced methane is dissolved in the liquor in anaerobic processes for treating municipal wastewater with a soluble COD of 200 mg/L at 30°C (Liu *et al.*, 2013). As such, recoverable methane is inevitably compromised. Another important challenge is to maintain a relatively high temperature in anaerobic processes, which is essential for the metabolisms of anaerobic microorganisms. It should also be noted that mainstream anaerobic processes for municipal wastewater treatment are unable to remove both ammonia and phosphate, thus further nutrients removal or recovery is definitely also needed.

2.3.2 Bio-electrochemical technologies

A bio-electrochemical system (BES) is an electrochemical cell combining microbial and electrochemical reactions, with electricity or other valuable outcomes (e.g. hydrogen or methane gas) as end products. Because of such unique features, BES has been considered as a promising technology for producing electrical energy, while also treating wastewater (Rozendal *et al.*, 2008). In general, a BES consists of an anode chamber and a cathode chamber, separated by an ion-exchange membrane. Through oxidation reactions in the anode chamber, generated protons diffuse through the ion exchange membrane, whereas produced electrons are directed to the cathode chamber to trigger reduction reactions. Currently, microbial fuel cells (MFCs) and microbial electrolysis cells (MECs) are the most commonly employed BESs for various applications (Logan & Rabaey, 2012).

2.3.2.1 *Microbial fuel cells*

In MFCs, electrical energy can be harvested due to the negative Gibbs free energy change of the overall reaction. As illustrated in Figure 2.6, organic matter in wastewater is degraded by electrochemically active microorganisms (EAMs) in

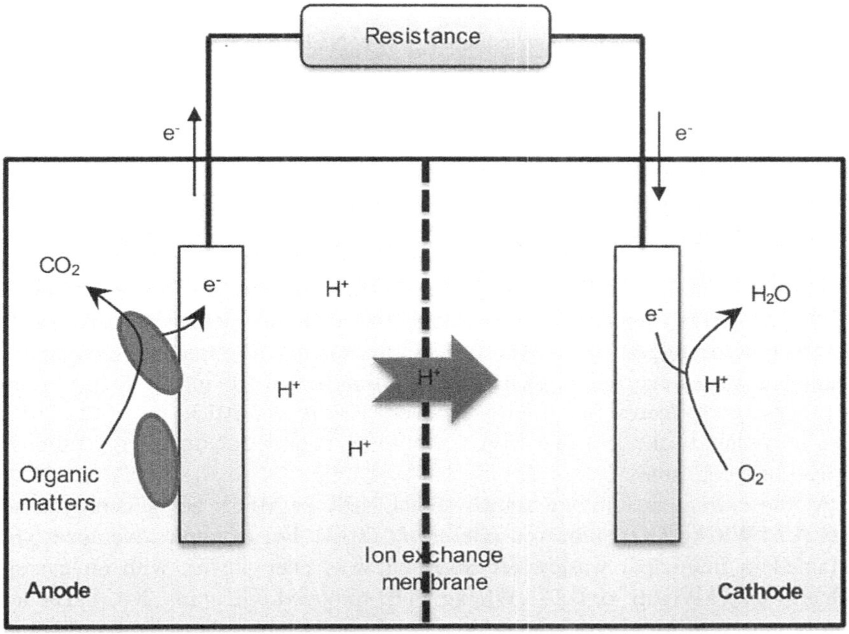

Figure 2.6 The working principles of MFCs. Rozendal *et al.* (2008), with permission.

the anode chamber, and produced electrons are ultimately transferred to the cathode, generating electrical current. Meanwhile, protons produced in the anode chamber tend to migrate to the cathode through a cation exchange membrane, where they combine with electrons, while reducing oxygen to water in air-breathing cathode MFCs (Rabaey *et al.*, 2009). Due to such unique features, MFCs may offer an engineering alternative for sustainable wastewater treatment, with the advantages of direct electricity generation, low sludge yield, modular design, small carbon footprint etc. (Li *et al.*, 2016). The sludge yield in MFCs had been reported to be about 0.1 g dry biomass/g COD removed (Zhang *et al.*, 2013), which is much lower than that in the CAS process with a typical value of 0.3–0.5 dry biomass/g COD removed. Moreover, MFCs also exhibit a small carbon footprint due to low energy consumption and carbon-neutral reactions, e.g. the oxidation of organic matter in the anode chamber only releases fixed carbon back into the atmosphere (Lovley, 2006).

Normalized energy recovery (NER) in kWh/m^3 wastewater treated or kWh/kg COD removed has been commonly used for evaluating MFCs performance (Ge *et al.*, 2013), and can also provide a baseline for cross-comparison with other wastewater treatment processes. After consideration of potential energy losses, including Coulombic loss, microbial assimilation and voltage loss, about 44% of influent COD could be ultimately converted to electrical energy (McCarty *et al.*,

2011). Therefore, for a typical municipal wastewater with a COD concentration of 500 mg/L, the recoverable electrical energy (i.e. NER) can be roughly estimated as 500 g/m^3 × 44% × 13.91 kJ/g = 3,060 kJ/m^3, i.e. 0.85 kWh/m^3 or 1.70 kWh/kg COD removed. In a laboratory study, even higher NER has also been reported, e.g. 2.2 kWh/m^3 or 2.0 kWh/kg COD removed (Ge *et al.*, 2013). These results suggest that the energy efficiency of MFCs could reach about 52% if the theoretical energy content in one kilogram of COD was assumed to be 3.86 kWh. Obviously, this energy efficiency is much higher than that obtained in the other anaerobic processes which are normally in the range of 13.7–25.6%. Moreover, the energy in MFCs are mainly utilized by pumps for feeding and recirculation, with an estimated energy demand of less than 0.05 kWh/m^3 (Zhang *et al.*, 2013). Evidence also suggests that MFCs are more suitable for treating low-strength municipal wastewater with a COD concentration below 500 mg COD/L in terms of process performance and energy recovery (Ge *et al.*, 2013; Yu *et al.*, 2012). This all points to the fact that MFCs should be regarded as an energy producer instead of a consumer.

At the current technology status, actual NER in MFCs are generally lower than 0.2 kWh/kg COD removed (Ge *et al.*, 2013). For example, averaged NER obtained in municipal wastewater treatment was even lower, with an average value of 0.04 kWh/m^3 or 0.17 kWh/kg COD removed (Ge *et al.*, 2013). The low energy output of MFCs is mainly due to the low pH buffer capacity and complex composition of municipal wastewater. In MFCs, substrates are metabolized to exchange electrons with the electrode by EAMs, but the EAM potential and the interactions between microbes and electrodes have not been fully understood. It should also be noted that high capital costs due to the use of expensive electrodes and separator materials is one of the barriers limiting the application of MFCs (Li *et al.*, 2014; Rozendal *et al.*, 2008). Although the performance of MFCs has, to some extent, been improved, their low power density is still the biggest challenge for their practical application. However, it is thought that high-performance MFCs could be used for municipal wastewater treatment with continuous efforts dedicated to their development. One such example is the power density of a polypyrrole-coated bacterial electrode which was increased by a factor of 14.1 compared to one without surface coating (Song *et al.*, 2015).

2.3.2.2 Microbial electrolysis cells

The Gibbs free energy change of the overall reaction is positive in microbial electrolysis cells (MECs), thus external energy is needed to drive the process, with the production of value-added products. The working mechanism of the anode in MECs is similar to that in an MFC (Figure 2.7). In the cathode chamber of MECs, protons are reduced to hydrogen gas through direct chemical catalysis (e.g. with platinum) or biocatalysis if a microbial bio-cathode is used, while utilizing electrons from the cathode in the absence of oxygen (Rabaey *et al.*, 2009).

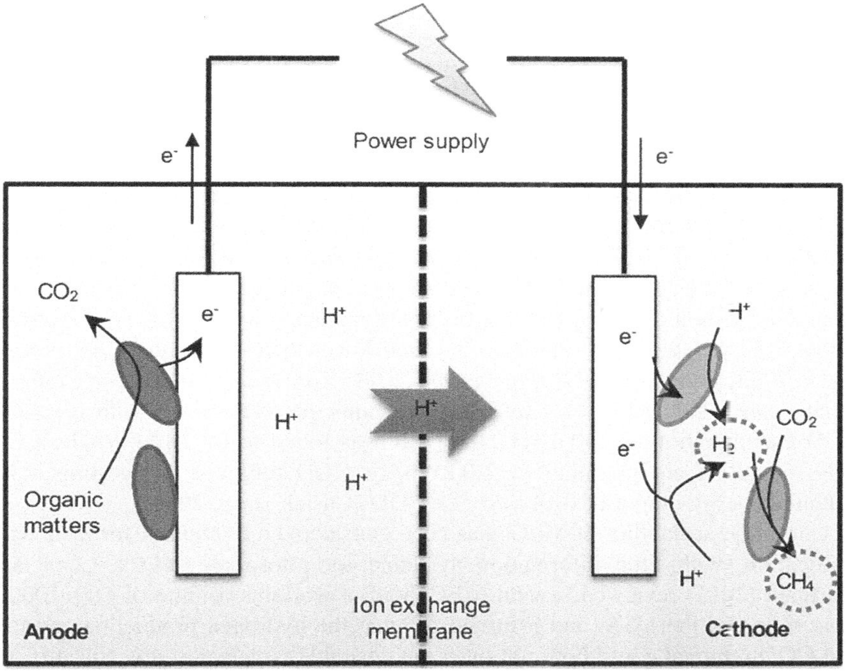

Figure 2.7 The working principles of MECs. Rozendal *et al.* (2008), with permission.

Hydrogen has been recognized as a green energy fuel with a high unit energy density, without carbon emission during its combustion. According to Equation 2.1 (Petrucci *et al.*, 2017):

$$H_2 + 0.5O_2 \rightarrow H_2O \qquad \Delta H_R^{\Theta} = -286 \text{ kJ/mol } H_2 \tag{2.1}$$

the chemical energy that can be released via combustion is theoretically estimated to be 11,690 kJ/m^3 of hydrogen gas (at 25°C under atmospheric pressure). Assuming an electrical conversion efficiency of 35%, about 1.14 kWh (11,690/3,600 × 35%) of electrical energy can be produced from one cubic meter of hydrogen gas. Hydrogen gas was the main product in a cathode chamber of MECs to which a small external voltage was applied (Logan *et al.*, 2008). For this purpose, a minimum voltage of − 0.13 V should be needed, which indeed is much lower than the theoretical voltage required for water splitting (e.g. about −1.2 V), due to the favourable thermodynamics of organic matter degradation compared to the highly energy-demanding reaction of water splitting (Logan & Rabaey, 2012). In practice, a higher negative voltage of above −0.3 V could be applied to accelerate hydrogen gas production.

The specific hydrogen gas production rate, in terms of m^3 $H_2/(m^3$ MEC · d), has been commonly used to evaluate the performance of MECs. Currently, specific hydrogen gas production rates in most MECs are below 5 m^3 $H_2/(m^3$ MEC · d), while the highest value of 50 m^3 $H_2/(m^3$ MEC · d) is reported at a higher electricity input of 2.5 kWh/m^3 H_2 (Jeremiasse *et al.*, 2011; Zou & He, 2018). In terms of NER in kWh/m^3 H_2 or kWh/kg COD, 1.0–3.1 kWh of electrical energy could be produced from one kilogram of COD with an energy input of 0.8–2.8 kWh/kg COD removed, indicating a net energy production of 0.2–0.3 kWh/kg COD removed in MECs. Although the energy efficiency of MECs (e.g. 5–8%) is much lower than those in MFCs and in the anaerobic process for wastewater treatment, it should be realized that the energy demand in an MEC (e.g. typically below 0.1 kWh/kg COD removed) is much lower than that in the CAS process (i.e. 0.7–2.0 kWh/kg COD) (Pant *et al.*, 2011; Zou *et al.*, 2017). For example, at lab-scale, using MECs for treating real municipal wastewater with a soluble COD concentration of 160 mg/L, the NER was found to be 2.14 kWh/kg COD removed at an energy input of 2.0 kWh/kg COD removed, suggesting a net potential energy output of 0.14 kWh/kg COD (Cusick *et al.*, 2010).

Currently, scalability of MECs has been considered a bottleneck for their wide application, with little information available for pilot-scale MECs. Lema and Martinez (2017) reviewed several MECs with a working volume of 100–1000 L in Europe and the USA, and it turned out that the hydrogen production capacity and COD removal could both not meet the desirable targets, i.e. it is still difficult to achieve energy self-sufficient wastewater treatment and good-quality effluent meeting discharge standards. This might be largely due to the fact that EAMs could not efficiently utilize particulate COD and long-chain organic matter in municipal wastewater. Therefore, a combined AD and MECs system may offer a great opportunity for maximizing the energy recovery from total organic matter including particulate COD (pCOD) and soluble COD (sCOD). On the other hand, the external power supply for hydrogen production accounts for a significant portion of energy demand at the system level (e.g. >95%) in MECs. Thus, the use of renewable power sources (e.g. solar energy) might provide an alternative solution to improve the overall energy efficiency of MECs. Apart from the hydrogen production in MECs, some value-added products (e.g. hydrogen peroxide, ethanol etc.) could also be harvested, which in turn could help to improve the MECs' overall economic viability (Pant *et al.*, 2011).

2.3.3 Energy recovery from nitrogenous compounds

Nitrogenous compounds in municipal wastewater are mainly present in the forms of organic nitrogen and ammonia. Currently, various biological nutrient removal processes have been developed, which are based on the philosophy of "eliminating," but at the costs of high energy consumption and generation of excess biosludge. In fact, the chemical energy embedded in nitrogenous compounds is considerable. As presented in Chapter 1, up to 0.3 kWh/m^3 of potential energy could be recovered from ammonia in municipal wastewater, implying that ammonia should not be simply removed through coupled bio-oxidation and bio-reduction (e.g. nitrification and denitrification) without considering energy recovery. To date, three nitrogenous compounds (i.e. ammonia, nitrous oxide and hydrazine) have been explored as potential energy sources recoverable from municipal wastewater.

2.3.3.1 Energy recovery from ammonia

Ammonia is considered to be a green energy fuel which produces nitrogen and water when combusted with oxygen (Equation 2.2), without carbon emission (Petrucci *et al.*, 2017):

$$NH_3 + 0.75O_2 \rightarrow 0.5N_2 + 1.5H_2O \qquad \Delta H_R^{\ominus} = -382\ \text{kJ/mol}\ NH_3 \tag{2.2}$$

Compared with hydrogen gas, ammonia, with a higher energy density, can be easily stored, transported and utilized as a liquid fuel in a safer manner. It has been demonstrated that ammonia gas can be liquefied at −10°C under a certain pressure, while the liquefaction of hydrogen gas requires an extremely low temperature of −253°C, or higher pressure (Gregory *et al.*, 1972). Increasing interest has been devoted to exploring ammonia-based clean fuel and ammonia economy (Service, 2018). For example, Australia is positioning itself to lead a renewable ammonia project with AU$12 million in 2018, while Japan is also strategically exploiting ammonia as a promising green fuel for fuel cell cars by 2030. It is likely that ammonia will become a clean fuel that helps to safeguard the environment with nearly zero carbon emission.

Ammonia (NH_3) and ammonium (NH_4^+) are conjugate pairs in water with a pK_a of 9.3 (Equation 2.3):

$$NH_3 + H^+ \rightarrow NH_4^+ \qquad pK_a = 9.3 \tag{2.3}$$

The pH of municipal wastewater is generally below 9.3 and thus ammonia mainly exists in the form of ammonium in water. This implies that the conversion of ammonium to free ammonia (FA) is prerequisite for ammonia recovery from wastewater. Currently, electrochemical fuel cells and physical-chemical methods have been explored for the harvesting of ammonia. According to the principle of charge neutrality, MFCs and MECs have been shown to have the capability of recovering ammonia from wastewater in the presence of a cation exchange membrane (Lema & Martinez, 2017). The electron transport in the external circuit forces the ammonium ion to migrate from the anode chamber to the cathode chamber through the cation exchange membrane. In the cathode chamber, the high pH of the catholyte favours the conversion of ammonium ions to highly volatile free ammonia which could be easily recovered through stripping or other approaches (Kuntke *et al.*, 2012; Zou & He, 2018). Gas stripping is based on the principle of the mass transfer by which air or other inert gas is used to strip ammonia gas out of wastewater. Since a higher pH is required for efficient ammonia stripping, lime is usually used to increase the wastewater pH (Kinidi *et al.*, 2018). In addition, electrodeionization (EDI) and solid oxide fuel cells (SOFCs) have also been explored for energy recovery from wastewater ammonia (Xu *et al.*, 2017).

The maximum energy potential of ammonia harvested from municipal wastewater having a concentration of 40 mg N/L has been estimated to be around 0.3 kWh/m^3. Supposing 35% for electricity conversion efficiency, the recoverable electrical energy from ammonia can be calculated as $0.3\ kWh/m^3 \times 35\% = 0.105\ kWh/m^3$, which may offset about 22.3% of the energy consumption in the CAS process with an average value of 0.47 kWh/m^3. However, it should be pointed out that the reported technologies for ammonia recovery were mainly developed for high-strength wastewater, e.g. digestate, leachate, industrial wastewater etc. In fact, the generation and harvesting of ammonia require the input of a substantial amount of energy, while post purification of harvested ammonia gas is needed to increase its combustion value. On the other hand, the toxicity of ammonia should be paid adequate attention. As an emerging technology, energy recovery from ammonia is still at the infancy stage, and its engineering feasibility and economic viability should be evaluated in a holistic manner.

2.3.3.2 Energy recovery from nitrous oxide

Nitrous oxide (N_2O) is known as a reactive nitrogenous compound containing a certain amount of chemical energy which can be released and recovered through thermal decomposition (Equation 2.4):

$$N_2O \rightarrow N_2 + 0.5O_2 \qquad \Delta H_R^\Theta = -82\ kJ/mol\ N_2O \tag{2.4}$$

In the conventional biological nitrogen removal process, nitrous oxide as a common intermediate can be generated via different metabolic pathways of nitrogen (Kampschreur *et al.*, 2009; Schreiber *et al.*, 2012), and it has usually been regarded as an unwanted byproduct due to its potent greenhouse effect and destruction of the ozone layer (Myhre *et al.*, 2013). In fact, nitrous oxide has been explored as a fuel additive in rocket propellants and is also popular for motor racing to enhance engine power output (Pfahl *et al.*, 2000). Moreover, nitrous oxide can also serve as a strong oxidizer or additive for co-combustion with other fuels (Powell *et al.*, 2008). Theoretically, according to Equations 2.5 and 2.6:

$$CH_4 + 4N_2O \rightarrow CO_2 + 2H_2O + 4N_2 \qquad \Delta H_R^\Theta = -1219\ kJ/mol\ CH_4 \tag{2.5}$$

$$CH_4 + 2O_2 \rightarrow CO_2 + 2H_2O \qquad \Delta H_R^\Theta = -890\ kJ/mol\ CH_4 \tag{2.6}$$

extra energy of 329 kJ/mol CH_4 could be generated through the co-combustion of nitrous oxide with methane compared to the combustion of methane with oxygen. As such, extensive efforts have been devoted to developing novel processes for enhanced nitrous oxide production from wastewater (Lin *et al.*, 2018; Myung *et al.*, 2015).

As illustrated in Figure 2.8, nitrous oxide can be generated through biological or coupled biological-chemical pathways in wastewater treatment (Zhang *et al.*, 2019). In biological nitrogen removal processes, nitrous oxide can be generated via three pathways, i.e. hydroxylamine (NH_2OH) oxidation, nitrifier denitrification, and denitrification. So far, pathways (1) and (2) have been commonly reported for nitrous oxide generation in biological processes, which are usually realized by ammonia-oxidizing bacteria (Ahn *et al.*, 2010; Law *et al.*, 2012). Nitrous oxide is also a common intermediate of denitrification by autotrophic or heterotrophic bacteria (Shapleigh, 2013). To improve the energy recovery from nitrous oxide, nitrate reduction should be blocked at the nitrous oxide step, while inhibiting its subsequent reduction to nitrogen gas. In addition, an iron-mediated reaction may offer a possibility of chemical generation of nitrous oxide (Moraghan & Buresh, 1977). However, it should be noted that the chemical pathways for generating nitrous oxide indeed rely on the supply of NH_2OH and NO_2^- produced via biological pathways (Figure 2.8).

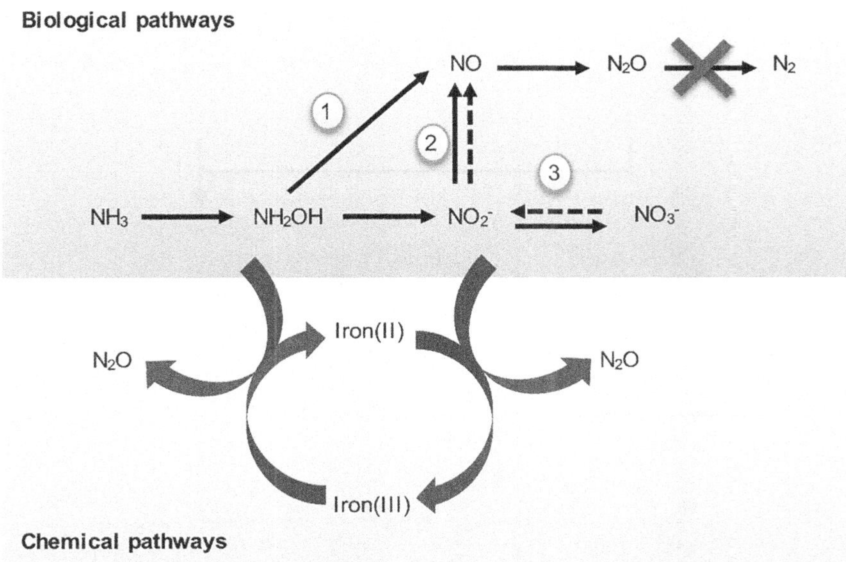

Figure 2.8 Pathways for N_2O generation in wastewater treatment processes. (1): Hydroxylamine (NH_2OH) oxidation; (2): Nitrifier denitrification; (3): Denitrification. Taken from Zhang *et al.* (2019), with permission.

For municipal wastewater with an ammonium concentration of 40 mg N/L, the produced nitrous oxide could be estimated as about 63 g/m^3 if all ammonium-N is converted to nitrous oxide. Given the unit chemical energy of nitrous oxide of 1.9 kJ/g (Equation 2.4), the maximum potential energy recoverable from nitrous oxide generated from municipal wastewater is calculated as 63 g/m^3 $\times$ 1.9 kJ/g = 119.7 kJ/m^3, i.e. 0.03 kWh/m^3 which is only about 6% of the in-plant energy consumption in the CAS process. In practice, the recoverable energy from nitrous oxide should be even smaller than this value due to the low conversion efficiency of chemical energy to electricity.

As mentioned above, the recovery of nitrous oxide from low-strength wastewater (e.g. municipal wastewater) may not be practically feasible and economically viable in consideration of the low unit recoverable energy in terms of kWh/m^3. By contrast, wastewater with high concentration ammonium (e.g. anaerobic digester liquor with an ammonium concentration of 500–1,500 mg N/L) may be more suitable for nitrous oxide production. As such, the Coupled Aerobic-anoxic Nitrous Decomposition Operation (CANDO) process was developed for the production and recovery of nitrous oxide from AD liquor (Scherson *et al.*, 2013). Typically, the CANDO is a three-step process, i.e. (1) partial nitrification of ammonia to nitrite; (2) partial denitrification (PDN) of nitrite to nitrous oxide; and (3) co-combustion of nitrous oxide with methane for energy generation (Figure 2.9).

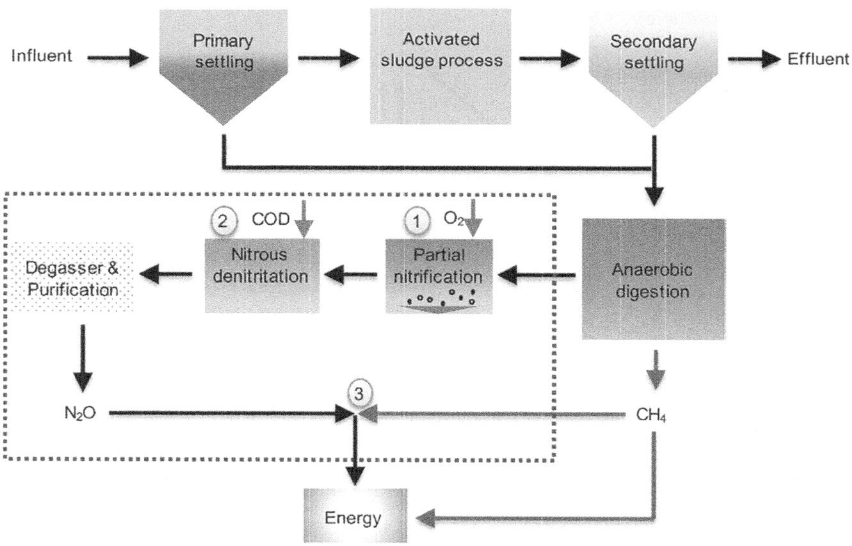

Figure 2.9 Wastewater treatment coupled with CANDO process. The dotted line box indicates the CANDO process: (1) partial nitrification of ammonia to nitrite; (2) nitrous denitritation of nitrite to nitrous oxide; and (3): co-combustion of nitrous oxide with methane. Zhang *et al.* (2019), with permission.

It has been reported that about 60–80% of nitrite could be converted to nitrous oxide in the CANDO process and the specific nitrous oxide generation rate would be as high as 25.2 mg N/(g VSS · h) (Myung *et al.*, 2015; Scherson *et al.*, 2014). Moreover, the CANDO process is also capable of simultaneously removing nitrogen and phosphorus through alternating anaerobic/anoxic cycles (Gao *et al.*, 2017).

Using the CANDO process to treat AD liquor with an ammonium concentration of 1,000 mg N/L, the generated nitrous oxide was estimated at 1,194.3 g N_2O/m^3 (Zhang *et al.*, 2019). Given a unit chemical energy of nitrous oxide of 1.9 kJ/g, the net recoverable energy from nitrous oxide could stand at 1.9 kJ/g $N_2O \times$ 1,194.3 g $N_2O/m^3 = 2{,}269.2$ kJ/m^3, equivalent to 0.22 kWh generated from one cubic meter of AD liquor through CHP at an electrical conversion efficiency of 35%. It should be noted that the organic carbon demanded for heterotrophic denitritation of nitrite to nitrous oxide would be reduced by 60% in the CANDO process compared to conventional complete denitrification in the CAS process, suggesting that saved organics can be directed to the anaerobic unit for methane production, which helps to partially offset the energy demand in WWTPs.

It should also be noted that anaerobic digester liquor only accounts for about 1% of the mainstream inflow in a WWTP (Fux & Siegrist, 2004), indicating that only about 0.22 kWh/m^3 anaerobic digester liquor $\times$ 1% = 0.0022 kWh/m^3 of wastewater treated could be generated, which accounts for about 0.5% of total energy consumption in WWTPs. Moreover, harvesting and further purification of the produced nitrous oxide are necessary prior to subsequent combustion for energy recovery, while the potent environmental risk related to the residual dissolved nitrous oxide should also be carefully assessed (Zhang *et al.*, 2019). Consequently, it appears reasonable to consider that the production and recovery technologies for nitrous oxide as a fuel are still at the early developmental stage, while further efforts will be needed to explore feasible engineering options for high-efficiency and cost-effective recovery of nitrous oxide with the aim to improving the overall energy efficiency of WWTPs.

2.3.3.3 Energy recovery from hydrazine

Hydrazine (N_2H_4) is known as a popular rocket fuel due to its high reducing power. Generally, it is a highly unusual metabolic intermediate because of its extremely high reactivity and low redox potential ($E'_0 = -750$ mV). The anaerobic ammonia oxidation (anammox) process has been attracting increasing attention for wastewater treatment due to its significantly reduced energy consumption and sludge production (Kartal *et al.*, 2010). Recent research has revealed that anammox bacteria are capable of producing hydrazine as an intermediate (Kartal *et al.*, 2011). An accumulation of 18 μmol/L of hydrazine was observed in an anammox system incubated with 0.1 mmol/L of nitric oxide and 2 mmol/L of ammonium (Kartal *et al.*, 2011).

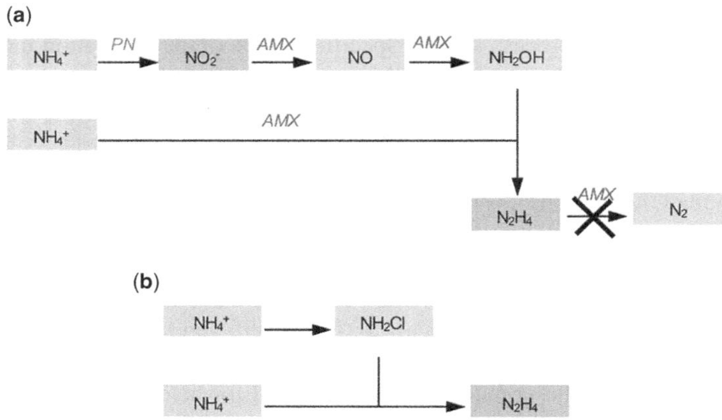

Figure 2.10 Hydrazine synthesis in anammox process for wastewater treatment (a) and industrial Olin-Raschig process (b). PN = partial nitrification; AMX = anammox.

In the anammox process for wastewater treatment, ammonium is first partially oxidized to nitrite by ammonia-oxidizing bacteria. As can be seen in Figure 2.10a, the produced nitrite further reacts with the remaining influent ammonium, with nitric oxide, hydroxylamine and hydrazine as intermediates (Dietl *et al.*, 2015). In fact, the proposed scheme for hydrazine synthesis is similar to the two-step industrial Olin-Raschig process developed in 1906, in which ammonium is oxidized to chloramine (NH_2Cl, like hydroxylamine) which reacts with another molecule of ammonium through comproportionation to yield hydrazine (Figure 2.10b). With a deeper understanding of how the hydrazine synthase functions, it would be possible to block further anammox reactions at the stage of hydrazine by inhibiting the protein activity. This provides a possible engineering means for maximizing the accumulation of hydrazine.

2.4 ENERGY-SAVING SHORT-CUT BIOLOGICAL NITROGEN REMOVAL PROCESSES

In the CAS process, ammonia is oxidized to nitrate with oxygen as the electron acceptor and is further reduced to nitrogen gas with consumption soluble COD as the electron donor (Tchobanoglous *et al.*, 2003). However, this biological nitrogen removal process is receiving a lot of criticism for the high aeration-associated energy consumption for nitrification, external COD demand for denitrification and the substantial production of waste sludge (Liu *et al.*, 2018). For example, the energy demand-associated with nitrification may account for about 21% of the total in-plant energy consumption (e.g. 0.47 kWh/m^3) (Goldstein & Smith, 2002; Monteith *et al.*, 2007). In general, soluble COD in municipal wastewater may not be enough to completely denitrify nitrate produced from nitrification. As such, additional organic carbon (e.g. methanol, ethanol, etc.) is needed to ensure complete denitrification, e.g. about 1.9–3.3 kg of methanol was required to denitrify one kilogram of nitrate-N in a full-scale WWTP (Theis & Hicks, 2012). Moreover, about 0.86 g VSS was produced in removing one gram of NH_4^+-N in a conventional biological nitrogen removal process with acetate as the carbon source (Ahn, 2006; Rittmann & McCarty, 2001), indicating that a non-negligible amount of waste sludge would inevitably be produced in the conventional biological nitrogen removal process. There is no doubt that the addition of external COD and post-treatment of waste sludge pushes up the operation costs and energy demand for biological nitrogen removal. In fact, the conventional biological nitrogen removal process has been identified as a big energy drain in WWTPs.

The processes presented in Section 2.2 are able to capture wastewater COD directly or indirectly. However, such COD capture may lead to insufficient COD for subsequent denitrification. According to the stoichiometric reaction of denitrification, about 2.86 gram of soluble COD is theoretically required to completely denitrify one gram of nitrate-N, suggesting that the conventional nitrification-denitrification process could not be feasible after prior capture of wastewater COD. Faced with such a situation, there is a growing need to explore short-cut nitrogen removal processes to move towards sustainable biological nitrogen removal.

2.4.1 Short-cut nitrogen removal processes

Conventional biological nitrogen removal is generally realized through complete nitrification and denitrification with multiple-step reactions, i.e. oxidation of ammonia to nitrite by ammonia oxidizing bacteria (AOB) (step 1), further oxidation of nitrite to nitrate by nitrite oxidizing bacteria (NOB) (step 2), followed by denitrification of nitrate to nitrite (step 3) and nitrogen gas (step 4) by denitrifying bacteria. So far, three typical short-cut processes, i.e. nitritation-denitritation, partial nitritation-anammox (PN-anammox) and partial denitrification-anammox (PDN-anammox) have been developed for energy-efficient nitrogen removal, as illustrated in Figure 2.11.

In the nitritation-denitritation process (steps 1 + 4), influent ammonia is only oxidized to nitrite under certain conditions by which NOB growth and activity are suppressed, followed by denitritation. In this process, the oxidation of nitrite to nitrate (step 2) and subsequent heterotrophic reduction of nitrate to nitrite (step 3) are eliminated, leading to reduced oxygen demand and COD requirement. As shown in Equations 2.7 and 2.8, 25% and 40% of reduction in dissolved oxygen and COD, respectively, are theoretically achievable:

$$NH_4^+ + 3/2O_2 \rightarrow NO_2^- + 2H^+ + H_2O \qquad (2.7)$$

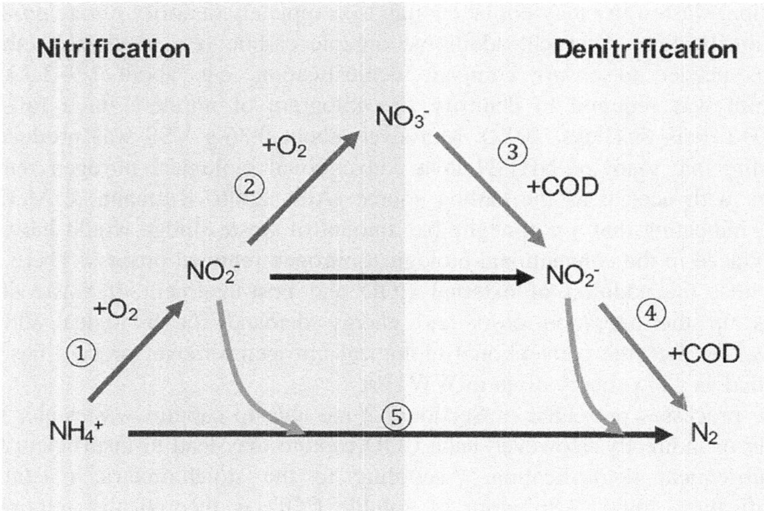

Figure 2.11 Short-cut processes for biological nitrogen removal. 1: partial nitrification (PN) or nitritation; 2: nitratation; 3: partial denitrification (PDN) or denitratation; 4: denitritation; 5: anammox.

$$NO_2^- + 3e^- + 2H^+ \rightarrow 1/2N_2 + H_2O \tag{2.8}$$

In the PN/anammox and PDN/anammox processes, the reduction of nitrite to nitrogen gas is carried out by anammox bacteria instead of by a denitrifier. Anaerobic ammonia oxidation known as anammox is an autotrophic biological process in which anammox bacteria use nitrite as the electron acceptor to oxidize ammonium to nitrogen gas (Equation 2.9) (Kartal *et al.*, 2010; Mulder *et al.*, 1995):

$$\begin{aligned} NH_4^+ + 1.32\,NO_2^- + 0.066HCO_3^- + 0.13H^+ \rightarrow 1.02N_2 \\ + 0.26NO_3^- + 0.066\,CH_2O_{0.5}N_{0.15} + 2.03H_2O \end{aligned} \tag{2.9}$$

The PN/anammox process can reduce oxygen demand and sludge production, while external COD is no longer needed.

In the completely autotrophic PN-anammox process (steps 1 + 5), about 57% of influent ammonia is converted to nitrite through PN, which will further react with the remaining ammonia through anammox, with nitrogen gas as the main end product. Theoretically, about 89% of wastewater ammonia is removable through PN-anammox.

In the PDN-anammox process (step 1 + 2 + 3 + 5), an anoxic tank is commonly used as a lead unit where nitrate produced by nitrification is converted to nitrite via PDN in the presence of influent COD. The produced nitrite will further react with part of the influent ammonia in the subsequent anammox unit, while the remaining ammonia is then oxidized to nitrate through complete nitrification. As such, both oxygen and COD demands can be reduced significantly. However, it should also be noted that the PDN-anammox process generally has a more complicated configuration than the PN-anammox process.

2.4.2 Process analysis

Intensive energy demands and substantial production of waste sludge have posed big challenges to current WWTPs in terms of economic viability and environmental sustainability. As presented in Chapter 1, aeration is regarded as the main energy consumer, accounting for about 50% of the total energy consumption in WWTPs (Cao, 2011; Goldstein & Smith, 2002), while post-treatment of waste sludge may account for about 25–65% of total operation costs (Liu & Tay, 2001). The bio-oxidation of COD in current WWTPs inevitably decreases the energy recovery potential from wastewater. For process evaluation, wastewater COD, energy consumption and biosludge generation should be taken into account in a holistic manner. As shown in Table 2.1, the complete nitrification-denitrification process is the least efficient process with the highest oxygen/COD demands and biomass production for the removal of one gram of ammonium-N. In this sense, the nitritation-denitritation process may help to reduce oxygen and COD demands by 26.8% and 40.2% respectively, while it may reduce biosludge by 35.0%. The PN-anammox process is the most efficient process, with significant reductions in oxygen and COD demand of 61.0% and 89.9%, respectively, while producing 75.0% less biosludge. As described in Section 2.3, the CANDO process with nitrous oxide recovery is also considered to be a kind of short-cut nitrogen removal technology due to the elimination of the reaction step from nitrous oxide to nitrogen gas in denitrification.

Table 2.1 Conventional biological nitrogen removal process versus short-cut nitrogen removal processes.

Process	O_2[a], g/g N	COD[b], g/g N	Biomass[c], g VSS/g N
Complete nitrification-denitrification	4.1	2.86	2.0
Nitritation-denitritation	3.0	1.71	1.3
PN-anammox[d]	1.6	0.29	0.5
PDN-anammox	2.1	0.73	0.5
CANDO[e]	3.0	1.14	0.9

[a]Data from Gao *et al.* (2014).
[b]Value calculated according to the stoichiometry of short-cut reactions.
[c]Value including biomass from ammonia oxidation and nitrite or nitrate reduction.
[d]The generated nitrate in the anammox process is supposed to be removed by denitrification.
[f]Heterotrophic CANDO process.

So far, the sidestream PN-anammox process has been developed for treating ammonia-rich wastewater (e.g. anaerobic digester liquor) with many successful applications worldwide (Lackner *et al.*, 2014). For example, in the Strass WWTP in Austria, PN-anammox was successfully applied for sidestream nitrogen removal, in which energy consumption was reduced by 44%. Moreover, sidestream deammonification does not utilize any organic carbon in term of COD, and thus the saved COD could be captured for direct biogas production through AD. However, mainstream PN-anammox for municipal wastewater with low-ammonia concentration has not yet been fully demonstrated at a large scale and further work is obviously needed in this direction.

2.5 RESOURCE RECOVERY FROM MUNICIPAL WASTEWATER

With technological advancements, municipal wastewater should not be viewed as "waste" but rather as a source of fresh water and resources. Indeed, this view is driving the shift in wastewater treatment from a "removing" to a "recovering" philosophy.

2.5.1 Water recovery: the primary goal of wastewater reclamation

Traditionally the primary objective of wastewater treatment has been to transform wastewater into water clean enough to be safely discharged into surface water bodies, without danger to human health or unacceptable damage to the natural environment. However, with rapid population growth and urbanization, water scarcity is becoming increasingly severe. For example, the global water demand will hit 6,906 km^3 by 2030, while water supply could potentially only satisfy about 60% of that demand (Boccaletti *et al.*, 2010). Moreover, the distribution of water sources is highly dependent on geographic location and on the season. To tackle this pressing issue, efforts have been dedicated to tapping water sources from wastewater reclamation, as well as from surface water, groundwater and seawater desalination. Today, most treated municipal wastewater is still discharged to receiving water bodies, which is indeed a waste of the water resource. In fact, treated wastewater should be regarded as a potential water resource for water recycling and reuse (Smith *et al.*, 2018). This may lead to reduced demand for natural freshwater. At a global scale, average municipal wastewater production stood at about 360 km^3/year in 2013 (Flörke *et al.*, 2013; Sato *et al.*, 2013), which was about 7.3% of the global water demand (i.e. 4,930 km^3). It has been projected that the amount of municipal wastewater will continue to rise in the years to come, making it a stable source for future water supply. Moreover, reclaimed water is near to the subsequent users in a city, which decreases the delivery costs and also avoids pollution of purified water during long-distance transport. Compared to seawater desalination, municipal wastewater reclamation is more economically viable, e.g. 0.7–1.2 kWh/m^3 for NEWater production from municipal wastewater against 3.9–4.3 kWh/m^3 for seawater desalination (Vincent *et al.*, 2014). In short, the recovery of high-grade water from municipal wastewater has been increasingly recognized as a sustainable water management strategy all over the world.

In Singapore, treated municipal wastewater is further purified to high-grade reclaimed water, known as NEWater which places Singapore towards a position of water sustainability (Lee & Tan, 2016). Currently, five NEWater plants are able to supply up to 40% of Singapore's daily water needs, while NEWater is expected to meet up to 55% of Singapore's future water demand by 2030 (PUB, 2019). In NEWater production, the secondary effluent from the biological treatment process of municipal wastewater is further reclaimed through a number of barriers, including microfiltration for particle removal, reverse osmosis to remove undesirable contaminants and ultraviolet disinfection to ensure complete eradication of any remaining organisms (Figure 2.12). Currently, the produced NEWater in Singapore could be utilized for either direct/non-potable use or indirect potable use. For direct potable use, NEWater could serve as a water source for various industrial demands which require high-grade water. In this way, a substantial reduction in tap water consumption by industry is realized. For indirect use, NEWater is discharged to surface water bodies (e.g. reservoirs) where, after naturalization, it is subsequently treated in a water treatment plant to produce drinking water.

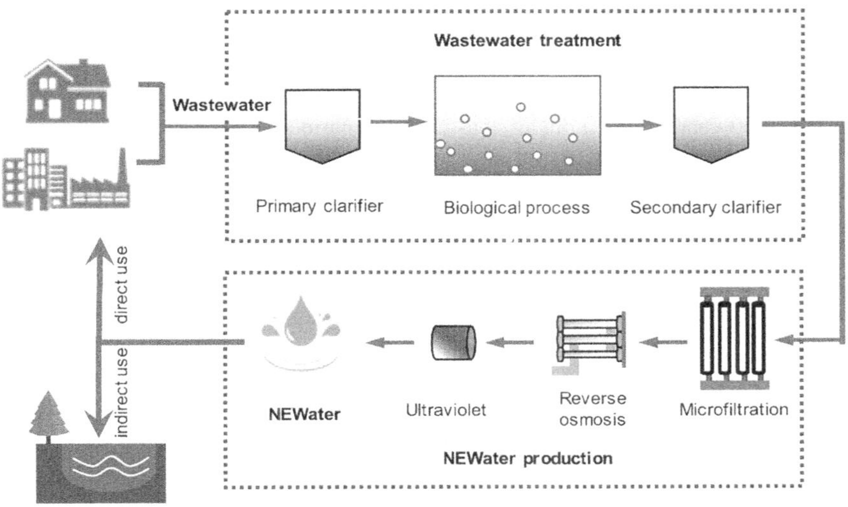

Figure 2.12 NEWater technology in Singapore. Information partly from PUB (2019).

In China, more and more cities suffer from increasing water shortage due to limited water resources. On the other hand, the annual treated municipal wastewater in China had already hit 46.7 billion (46.7×10^9) m^3 in 2015 (Smith *et al.*, 2018), and nearly 4,000 municipal WWTPs are currently in operation. Given this, reuse of reclaimed wastewater is a growing trend in order to satisfy increasing water demand (Chang *et al.*, 2013). For example, there were 571 water reclamation plants (WRPs) treating secondary effluents from conventional WWTPs for water reuse in 2013, with a total length of reclaimed water pipelines of about 8,300 km (Chen *et al.*, 2017). Reclaimed water was reported to increase from 8.2% to 9.5% of annually produced wastewater between 2014 and 2015 (Smith *et al.*, 2018). Table 2.2 shows the Chinese standards for reclaimed water organized in a hierarchy of different uses. To meet such stringent standards, various treatment technologies have been adopted in Chinese wastewater reuse plants. Moreover, it should be noted that the quality of reclaimed water in China is of hierarchical dependence, which is significantly different from NEWater. For example, it has been suggested that RO permeate could be blended with the treated effluent from the combined ultrafiltration-ozonation process at a ratio of 2:3 to achieve a suitable water quality for industrial reuse with a reduced treatment cost (Chen *et al.*, 2017). Moving forward, future WWTPs should be designed and operated as a water factory aiming to produce high-grade reclaimed water at an affordable cost.

2.5.2 Nutrient recovery: a necessity for environmental sustainability

Ammonia is a typical fertilizer commonly used in agriculture. So far, ammonia has mainly been produced through the well-known Haber-Bosch process which is energy-intensive and has a large carbon footprint. Phosphorus as a finite resource is non-renewable on a human timescale and is estimated to be exhausted in the next 50–400 years under a business as usual scenario (Reijnders, 2014). Currently, over 80% of annually produced phosphorus from mined fossil resources is used for fertilizer production. By contrast, it should be realized that a significant amount of ammonia and phosphorus are present in municipal wastewater, e.g. 40 mg/L of ammonium-N and 6–15 mg/L of phosphorus-P. Given the global municipal wastewater production of around 360 km^3 per year, the total amount of ammonium-N present in municipal wastewater is estimated to be 360 km^3/year $\times$ 40 mg/L = 14.4 million tonnes/year, accounting for 13% of the global demand for ammonia-based fertilizer, i.e. about 110 million in 2015 (FAO, 2017). Similarly, with an average phosphorus concentration of 10.5 mg/L, the total amount of phosphorus in municipal wastewater stood at 360 km^3/year $\times$ 10.5 mg/L = 3.78 million tonnes P/year, equivalent to 8.66 tonnes P_2O_5/year, which was 21% of the global P-fertilizer demand (i.e. 41.2 million tonnes P_2O_5/year). Obviously, municipal wastewater is indeed a source rich of nutrients. Therefore, a

Table 2.2 Chinese standards of reclaimed water for different uses (Lyu *et al.*, 2016).

Values	Industrial use		Agricultural irrigation			Non-potable Water Use		
Indexes, mg/L	**Cooling**	**Washing**	**Paddy Grain**	**Field Vegetables**	**Dryland Grain and Oil Crop**	**Toilet Flushing**	**Urban Greening**	**Construction**
pH	6.5–8.5	6.5–9.0	5.5–8.5	5.5–8.5	5.5–8.5	6.0–9.0	6.0–9.0	6.0–9.0
BOD_5	10	30	60	40	80	10	20	15
COD_{cr}	60	60	150	100	180	–	–	–
NH_3-N	10	10	–	–	–	10	20	20
TDS[a]	1,000	1,000	1,000/2,000[b]	1,000	1,000/2,000[b]	1,500	1,000	1,500
SS	30	30	80	60	90	–	–	–
Fe	0.3	0.3	1.5	1.5	1.5	0.3	1.5	–
Mn	0.1	0.1	0.3	0.3	0.3	0.1	0.3	–
Cr	–		0.1	0.1	0.1	–	0.1	–
Pb	–		0.2	0.2	0.2	–	0.2	–
As	–		0.05	0.05	0.1	–	0.05	–
Hg	–		0.001	0.001	0.001	–	0.001	–
Cd	–		0.01	0.01	0.01	–	0.01	–

[a]TDS = total dissolved solids.
[b]Smaller value for areas with non-saline soils/larger value for areas with saline soils.

paradigm shift in WWTPs from wastewater treatment to resource recovery is absolutely necessary.

Chemical precipitation is one of the most popular methods for recovering phosphorus from wastewater in the forms of struvite ($MgNH_4PO_4 \cdot 6H_2O$), hydroxyapatite ($Ca_5(PO_4)_3OH$), vivianite ($Fe_3(PO_4)_2 \cdot 8H_2O$) etc. By formation of struvite, both ammonia and phosphate can be simultaneously recovered as described in Equation 2.9 (Campos *et al.*, 2018), in which pH and magnesium dosing are the decisive factors. In general, the magnesium content in municipal wastewater cannot meet the requirement for struvite formation, i.e. additional dosage of magnesium is needed, leading to an increased production cost. In fact, uses of alkali and magnesium account for up to 75% of the overall production cost of struvite (Dockhorn, 2009). Moreover, it should be noted that the molar ratio of ammonia-N to phosphorus-P in municipal wastewater is much higher than that required for struvite formation (Equation 2.10), i.e. residual ammonia still needs to be handled after the struvite precipitation.

$$NH_4^+ + Mg^{2+} + PO_4^{3-} + 6H_2O \rightarrow MgNH_4PO_4 \cdot 6H_2O \tag{2.10}$$

Phosphate in wastewater can be recovered in the form of hydroxyapatite precipitate (Equation 2.11) by adjusting pH and dosing of calcium:

$$5Ca^{2+} + 3PO_4^{3-} + OH^- \rightarrow Ca_5(PO_4)_3OH \tag{2.11}$$

Similar to struvite, hydroxyapatite can be used as a fertilizer for agriculture. In addition, recovery of phosphorus in the form of vivianite (Equation 2.12) has also been explored as an additional option for phosphorus recovery from municipal wastewater (Wang *et al.*, 2019; Wilfert *et al.*, 2016):

$$3Fe^{2+} + 2PO_4^{3-} + 8H_2O \rightarrow Fe_3(PO_4)_2 \cdot 8H_2O \tag{2.12}$$

However, the technology for vivianite formation is still at its developmental stage. Further effort is strongly needed to improve its performance, engineering feasibility and economic viability.

Currently, there is an increasing interest in the AnMBR process for municipal wastewater treatment. An AnMBR is able to efficiently remove COD, while producing solids-free permeate rich in ammonia and phosphorus. In fact, it has been suggested that the combination of an AnMBR and nutrient recovery via direct or indirect methods may offer a promising option for municipal wastewater reclamation (Figure 2.13) (Liu *et al.*, 2018). However, almost all the present technologies developed for recovering nutrients (e.g. N and P) from municipal wastewater appear to be challenging in terms of economic viability and environmental sustainability (Liu *et al.*, 2018).

Furthermore, there is still a debate on the necessity of reclaiming all wastewater to the grade suitable for drinking. In fact, wastewater reclamation

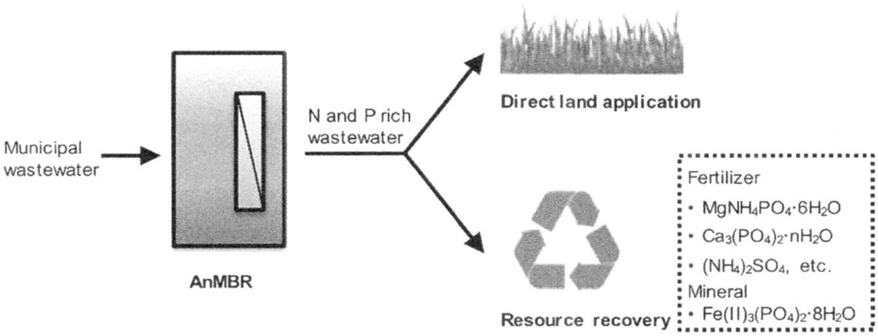

Figure 2.13 Nutrient recovery from municipal wastewater. Adapted from Liu *et al.* (2018), with permission.

should be task-driven in certain circumstances, which might be more technologically feasible and economically effective than converting all water to a potable grade (Tao & Xin, 2014). It is to be applauded that, for regions with varied requirements for water demand, the water policy should be flexible enough to make different grades of water accessible according to real needs. For example, in the Regional Water Pollution Control Plant (Monterey, USA), located in a prime vegetable producing but water-short agricultural area, anaerobically reclaimed water rich in N and P with a capacity of 76,000 m^3/day was reported to be available for irrigating 4,900 ha of vegetable crops (McCarty *et al.*, 2011).

2.6 CONCLUSIONS

Moving forward, municipal wastewater should not be viewed as a "waste", but as a source rich in various resources, while the ultimate goal of municipal wastewater treatment should move away from the current simple removal philosophy to the synergetic recovery of water, energy and resources (Figure 2.14). Toward this end, more efforts should be devoted to developing innovative technology platforms to more efficiently capture wastewater COD, e.g. the direct conversion of soluble COD to bio-methane via anaerobic treatment and channeling of more wastewater COD into biosludge for further AD. In addition to energy recovery in methane gas, increasing interest has been given to the energy recovery from nitrogenous compounds in wastewater in the forms of ammonia, nitrous oxide and hydrazine. Also, energy-saving biological nitrogen removal processes (e.g. nitritation-denitritation, partial nitritation-anammox and PDN-anammox) coupled with COD capture have shown great potential towards sustainable wastewater reclamation.

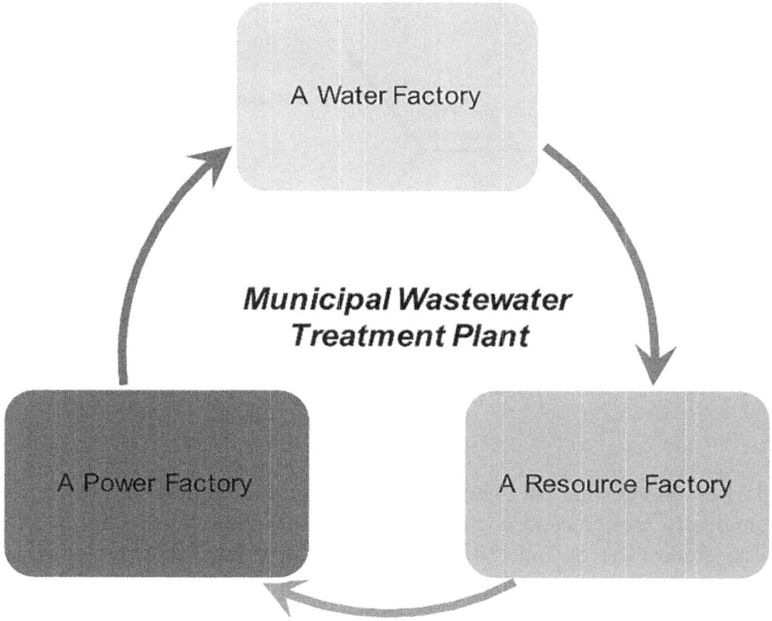

Figure 2.14 Municipal wastewater treatment plant: a source of water, energy and resources.

REFERENCES

Ahn J. H., Kim S., Park H., Rahm B., Pagilla K. and Chandran K. (2010). N_2O emissions from activated sludge processes, 2008–2009: results of a national monitoring survey in the United States. *Environmental Science & Technology*, **44**(12), 4505–4511.

Ahn Y. H. (2006). Sustainable nitrogen elimination biotechnologies: a review. *Process Biochemistry*, **41**(8), 1709–1721.

Barjenbruch M. and Kopplow O. (2003). Enzymatic, mechanical and thermal pre-treatment of surplus sludge. *Advances in Environmental Research*, **7**(3), 715–720.

Boccaletti G., Grobbel M. and Stuchtey M. R. (2010). The business opportunity in water conservation. *McKinsey Quarterly*, **1**, 67–75.

Boehnke B., Schulze-Rettmer R. and Zuckut S. (1998). Cost-effective reduction of high-strength wastewater by adsorption-based activated sludge technology. *Water Engineering & Management*, **145**(12), 31–34.

Bougrier C., Battimelli A., Delgenes J.-P. and Carrere H. (2007). Combined ozone pretreatment and anaerobic digestion for the reduction of biological sludge production in wastewater treatment. *Ozone: Science and Engineering*, **29**(3), 201–206.

Campos J. L., Crutchik D., Franchi Ó., Pavissich J. P., Belmonte M., Pedrouso A., Mosquera-Corral A. and Val del Río Á. (2018). Nitrogen and phosphorus recovery from anaerobically preteated agro-food wastes: A review. *Frontiers in Sustainable Food Systems*, **2**, 91.

Cao Y. S. (2011). Mass Flow and Energy Efficiency of Municipal Wastewater Treatment Plants. IWA Publishing, London, UK.

Carrère H., Dumas C., Battimelli A., Batstone D., Delgenès J., Steyer J. and Ferrer I. (2010). Pretreatment methods to improve sludge anaerobic degradability: a review. *Journal of Hazardous Materials*, **183**(1–3), 1–15.

Chang D., Ma Z. and Wang X. (2013). Framework of wastewater reclamation and reuse policies (WRRPs) in China: comparative analysis across levels and areas. *Environmental Science & Policy*, **33**, 41–52.

Chagnon F. and Harleman D. R. (2002). An Introduction to Chemically Enhanced Primary Treatment. Massachusetts Institute of Technology, Cambridge, MA.

Chen Z., Wu Q., Wu G. and Hu H.-Y. (2017). Centralized water reuse system with multiple applications in urban areas: Lessons from China's experience. *Resources, Conservation and Recycling*, **117**, 125–136.

Cusick R. D., Kiely P. D. and Logan B. E. (2010). A monetary comparison of energy recovered from microbial fuel cells and microbial electrolysis cells fed winery or domestic wastewaters. *International Journal of Hydrogen Energy*, **35**(17), 8855–8861.

Dai R., Liu Y., Liu X., Zhang X., Zeng C. and Li L. (2011). Investigation of a sewage-integrated technology combining an expanded granular sludge bed (EGSB) and an electrochemical reactor in a pilot-scale plant. *Journal of Hazardous Materials*, **192**(3), 1161–1170.

Diamantis V., Melidis P., Aivasidis A., Verstraete W. and Vlaeminck S. (2011). Efficiency and sustainability of urban wastewater treatment with maximum separation of the solid and liquid fraction. In: M. Moo-Young (ed.), (Volume 6): Environmental Biotechnology and Safety. Elsevier, pp. 507–515.

Diamantis V., Verstraete W., Eftaxias A., Bundervoet B., Siegfried V., Melidis P. and Aivasidis A. (2013). Sewage pre-concentration for maximum recovery and reuse at decentralized level. *Water Science and Technology*, **67**(6), 1188–1193.

Dietl A., Ferousi C., Maalcke W. J., Menzel A., de Vries S., Keltjens J. T., Jetten M. S., Kartal B. and Barends T. R. (2015). The inner workings of the hydrazine synthase multiprotein complex. *Nature*, **527**(7578), 394.

Dockhorn T. (2009). About the economy of phosphorus recovery. International Conference on Nutrient Recovery from Wastewater Streams. IWA Publishing, London, UK. pp. 145–158.

Dolejs P., Gotvald R., Velazquez A. M., Hejnic J., Jenicek P. and Bartacek J. (2016). Contact stabilization with enhanced accumulation process for energy recovery from sewage. *Environmental Engineering Science*, **33**(11), 873–881.

Edwards J., Othman M., Crossin E. and Burn S. (2017). Anaerobic co-digestion of municipal food waste and sewage sludge: A comparative life cycle assessment in the context of a waste service provision. *Bioresource Technology*, **223**, 237–249.

Elliott M. S. (2016). Impacts of Operating Parameters on Extracellular Polymeric Substances Production in a High Rate Activated Sludge System with Low Solids Retention Times. Master thesis, Old Dominion University, Virginia, USA.

Engelhart M., Krger M., Kopp J. and Dichtl N. (2000). Effects of disintegration on anaerobic degradation of sewage excess sludge in downflow stationary fixed film digesters. *Water Science and Technology*, **41**(3), 171–179.

Food and Agriculture Organization (FAO). (2017). World Fertilizer Trends and Outlook to 2020. FAO of the United Nations, Rome, summary report.

Florencio L., Kato M. T. and Cardoso de Morais J. (2001). Domestic sewage treatment in full-scale UASBB plant at Mangueira, Recife, Pernambuco. *Water Science and Technology*, **44**(4), 71–77.

Flörke M., Kynast E., Bärlund I., Eisner S., Wimmer F. and Alcamo J. (2013). Domestic and industrial water uses of the past 60 years as a mirror of socio-economic development: A global simulation study. *Global Environmental Change*, **23**(1), 144–156.

Fux C. and Siegrist H. (2004). Nitrogen removal from sludge digester liquids by nitrification/denitrification or partial nitritation/anammox: environmental and economical considerations. *Water Science and Technology*, **50**(10), 19–26.

Gao H., Scherson Y. D. and Wells G. F. (2014). Towards energy neutral wastewater treatment: methodology and state of the art. *Environmental Science: Processes & Impacts*, **16**(6), 1223–1246.

Gao H., Liu M., Griffin J. S., Xu L., Xiang D., Scherson Y. D., Liu W. T. and Wells G. F. (2017). Complete nutrient removal coupled to nitrous oxide production as a bioenergy source by denitrifying polyphosphate-accumulating organisms. *Environmental Science & Technology*, **51**(8), 4531–4540.

Ge Z., Li J., Xiao L., Tong Y. and He Z. (2013). Recovery of electrical energy in microbial fuel cells: brief review. *Environmental Science & Technology Letters*, **1**(2), 137–141.

Goldstein R. and Smith W. (2002). Water & Sustainability (volume 4): US Electricity Consumption for Water Supply & Treatment-The Next Half Century. Electric Power Research Institute, California, USA.

Gonzalez A., Hendriks A., Van Lier J. and de Kreuk M. (2018). Pre-treatments to enhance the biodegradability of waste activated sludge: elucidating the rate limiting step. *Biotechnology Advances*, **36**(5), 1434–1469.

González J., Sánchez M. and Gómez X. (2018). Enhancing anaerobic digestion: the effect of carbon conductive materials. *C*, **4**(4), 59.

Gottardo M., Micoluccib F., Mattiolib A., Faggiana S., Cavinatoa C. and Pavana P. (2015). Hydrogen and methane production from biowaste and sewage sludge by two phases anaerobic codigestion. *Chemical Engineering Transactions*, **43**, 379–384.

Gregory D. P., Ng D. and Long G. (1972). The hydrogen economy. In: J. O'M. Bockris (ed.), *Electrochemistry of Cleaner Environments*, Springer, New York, USA, pp. 226–280.

Gu J., Yang Q. and Liu Y. (2018). Mainstream anammox in a novel A-2B process for energy-efficient municipal wastewater treatment with minimized sludge production. *Water Research*, **138**, 1–6.

Heidrich E., Curtis T. and Dolfing J. (2010). Determination of the internal chemical energy of wastewater. *Environmental Science & Technology*, **45**(2), 827–832.

Henze M., van Loosdrecht M. C., Ekama G. A. and Brdjanovic D. (2008). *Biological Wastewater Treatment*. IWA Publishing, London, UK.

Huang J. C. and Li L. (2000). An innovative approach to maximize primary treatment performance. *Water Science and Technology*, **42**(12), 209–222.

Jeremiasse A. W., Bergsma J., Kleijn J. M., Saakes M., Buisman C. J., Stuart M. C. and Hamelers H. V. (2011). Performance of metal alloys as hydrogen evolution reaction catalysts in a microbial electrolysis cell. *International Journal of Hydrogen Energy*, **36**(17), 10482–10489.

Judd S. and Judd C. I. (2010). The MBR Book: Principles and Applications of Membrane Bioreactors for Water and Wastewater Treatment. Elsevier, Oxford, UK.

Kampschreur M. J., Temmink H., Kleerebezem R., Jetten M. S.M. and van Loosdrecht M. C. M. (2009). Nitrous oxide emission during wastewater treatment. *Water Research*, **43**(17), 4093–4103.

Kartal B., Kuenen J. V. and Van Loosdrecht M. (2010). Sewage treatment with anammox. *Science*, **328**(5979), 702–703.

Kartal B., Maalcke W., de Almeida N., Cirpus I., Gloerich J., Geerts W., den Camp H., Harhangi H., Janssen-Megens E., Francoijs K., Stunnenberg H., Keltjens J., Jetten M. and Strous M. (2011). Molecular mechanism of anaerobic ammonium oxidation. *Nature*, **479**(7371), 127–U159.

Kinidi L., Tan I. A.W., Wahab A., Binti N., Tamrin K. F.B., Hipolito C. N. and Salleh S. F. (2018). Recent development in ammonia stripping process for industrial wastewater treatment. *International Journal of Chemical Engineering*, **2018**.

Koch K., Plabst M., Schmidt A., Helmreich B. and Drewes J. E. (2016). Co-digestion of food waste in a municipal wastewater treatment plant: comparison of batch tests and full-scale experiences. *Waste Management*, **47**, 28–33.

Kuntke P., Śmiech K. M., Bruning H., Zeeman G., Saakes M., Sleutels T. H. J. A., Hamelers H. V.M. and Buisman C. J. N. (2012). Ammonium recovery and energy production from urine by a microbial fuel cell. *Water Research*, **46**(8), 2627–2636.

Lackner S., Gilbert E. M., Vlaeminck S. E., Joss A., Horn H. and van Loosdrecht M. C. (2014). Full-scale partial nitritation/anammox experiences–an application survey. *Water Research*, **55**, 292–303.

Law Y., Ye L., Pan Y. and Yuan Z. (2012). Nitrous oxide emissions from wastewater treatment processes. *Philosophical Transactions of the Royal Society B: Biological Sciences*, **367**(1593), 1265–1277.

Lee H. and Tan T. P. (2016). Singapore's experience with reclaimed water: NEWater. *International Journal of Water Resources Development*, **32**(4), 611–621.

Lema J. M. and Martinez S. S. (2017). *Innovative Wastewater Treatment & Resource Recovery Technologies: Impacts on Energy, Economy and Environment*. IWA publishing, London, UK.

Li D., Tan Y., Zhou Y., Pathak S., Sendjaja A. Y., Abdul Majid M., Chowdhury P. and Ng W. J. (2015). Comparative study of low-energy ultrasonic and alkaline treatment on biosludge from secondary industrial wastewater treatment. *Environmental Technology*, **36**(17), 2239–2248.

Li W.-W., Yu H.-Q. and He Z. (2014). Towards sustainable wastewater treatment by using microbial fuel cells-centered technologies. *Energy & Environmental Science*, **7**(3), 911–924.

Li X. J., Sun S., Badgley B. D., Sung S. W., Zhang H. S. and He Z. (2016). Nitrogen removal by granular nitritation-anammox in an upflow membrane-aerated biofilm reactor. *Water Research*, **94**, 23–31.

Li Y.-Y. and Noike T. (1992). Upgrading of anaerobic digestion of waste activated sludge by thermal pretreatment. *Water Science & Technology*, **26**(3–4), 857–866.

Lin Z., Sun D., Dang Y. and Holmes D. E. (2018). Significant enhancement of nitrous oxide energy yields from wastewater achieved by bioaugmentation with a recombinant strain of *Pseudomonas aeruginosa*. Scientific Reports, 8.

Liu S.-G., Ni B.-J., Wel L., Tang Y. and Yu H.-Q. (2009). Contact-adsorption-regeneration-stabilization process for the treatment of municipal wastewater. *Journal of Water and Environment Technology*, **7**(2), 83–90.

Liu Y. and Tay J.-H. (2001). Strategy for minimization of excess sludge production from the activated sludge process. *Biotechnology Advances*, **191**(2), 97–107.

Liu Y., Gu J. and Liu Y. (2018). Energy self-sufficient biological municipal wastewater reclamation: present status, challenges and solutions forward. *Bioresource Technology*, **269**, 513–519.

Liu Z., Yin H., Dang Z. and Liu Y. (2013). Dissolved methane: a hurdle for anaerobic treatment of municipal wastewater. *Environmental Science & Technology*, **48**(2), 889–890.

Logan B. E. and Rabaey K. (2012). Conversion of wastes into bioelectricity and chemicals by using microbial electrochemical technologies. *Science*, **337**(6095), 686–690.

Logan B. E., Call D., Cheng S., Hamelers H. V. M., Sleutels T. H. J. A., Jeremiasse A. W. and Rozendal R. A. (2008). Microbial electrolysis cells for high yield hydrogen gas production from organic matter. *Environmental Science & Technology*, **42**(23), 8630–8640.

Lovley D. R. (2006). Bug juice: harvesting electricity with microorganisms. *Nature Reviews Microbiology*, **4**(7), 497.

Lyu S., Chen W., Zhang W., Fan Y. and Jiao W. (2016). Wastewater reclamation and reuse in China: Opportunities and challenges. *Journal of Environmental Sciences*, **39**, 86–96.

Ma Y., Yin Y. and Liu Y. (2017). New insights into co-digestion of activated sludge and food waste: Biogas versus biofertilizer. *Bioresource Technology*, **241**, 448–453.

Ma Y., Gu J. and Liu Y. (2018). Evaluation of anaerobic digestion of food waste and waste activated sludge: Soluble COD versus its chemical composition. *Science of The Total Environment*, **643**, 21–27.

Maspolim Y., Guo C., Xiao K., Zhou Y. and Ng W. J. (2016). Performance and microbial community analysis in alkaline two-stage enhanced anaerobic sludge digestion system. *Biochemical Engineering Journal*, 105, 296–305.

McCarty P. L., Bae J. and Kim J. (2011). Domestic wastewater treatment as a net energy producer–can this be achieved? *Environmental Science & Technology*, **45**, 7100–7106.

Meerburg F. A., Boon N., Van Winckel T., Vercamer J. A.R., Nopens I. and Vlaeminck S. E. (2015). Toward energy-neutral wastewater treatment: A high-rate contact stabilization process to maximally recover sewage organics. *Bioresource Technology*, **179**, 373–381.

Metcalf E. and Eddy M. (2014). Wastewater Engineering: Treatment and Resource Recovery. Mic Graw-Hill, USA, 1530–1533.

Mihelcic J. R., Zimmerman J. B. and Auer M. T. (2014). Environmental Engineering: Fundamentals, Sustainability, Design. Wiley Hoboken, New Jersey, USA.

Miron Y., Zeeman G., Van Lier J. B. and Lettinga G. (2000). The role of sludge retention time in the hydrolysis and acidification of lipids, carbohydrates and proteins during digestion of primary sludge in CSTR systems. *Water Research*, **34**(5), 1705–1713.

Modin O., Persson F., Wilén B.-M. and Hermansson M. (2016). Nonoxidative removal of organics in the activated sludge process. *Critical Reviews in Environmental Science and Technology*, **46**(7), 635–672.

Monteith H., Kalogo Y. and Louzeiro N. (2007). Achieving stringent effluent limits takes a lot of energy! *Proceedings of the Water Environment Federation*, **2007**(14), 4343–4356.

Moraghan J. and Buresh R. (1977). Chemical reduction of nitrite and nitrous oxide by ferrous iron. *Soil Science Society of America Journal*, **41**(1), 47–50.

Mulder A., Vandergraaf A., Robertson L. and Kuenen J. (1995). Anaerobic ammonium oxidation discoveryed in a denitrifying fluidized-bed reactor. *Fems Microbiology Ecology*, **16**(3), 177–183.

Myhre G., Shindell D., Bréon F. M., Collins W., Fuglestvedt J., Huang J., Koch D., Lamarque J. F., Lee D. and Mendoza B. (2013). Anthropogenic and natural radiative forcing. *Climate Change*, **423**, 658–740.

Myung J., Wang Z., Yuan T., Zhang P., Van Nostrand J. D., Zhou J. and Criddle C. S. (2015). Production of nitrous oxide from nitrite in stable Type II methanotrophic enrichments. *Environmental Science & Technology*, **49**(18), 10969–10975.

Pant D., Singh A., Van Bogaert G., Gallego Y. A., Diels L. and Vanbroekhoven K. (2011). An introduction to the life cycle assessment (LCA) of bioelectrochemical systems (BES) for sustainable energy and product generation: relevance and key aspects. *Renewable and Sustainable Energy Reviews*, **15**(2), 1305–1313.

Parkin G. F. and Owen W. F. (1986). Fundamentals of anaerobic digestion of wastewater sludges. *Journal of Environmental Engineering*, **112**(5), 867–920.

Petrucci R. H., Herring F. G., Bissonnette C. and Madura J. D. (2017). General Chemistry: Principles and Modern Applications. Pearson, Toronto, Canada.

Pfahl U. J., Ross M. C., Shepherd J. E., Pasamehmetoglu K. O. and Unal C. (2000). Flammability limits, ignition energy, and flame speeds in H_2–CH_4–NH_3–N_2O–O_2–N_2 mixtures. *Combustion and Flame*, **123**(1), 140–158.

Powell O., Miller J., Dreyer C. and Papas P. (2008). Characterization of hydrocarbon/nitrous oxide propellant combinations. 46th AIAA Aerospace Sciences Meeting and Exhibit, p. 999.

PUB. (2019). NEWater Usage. https://www.pub.gov.sg/watersupply/fournationaltaps/newater (accessed 22nd January 2019).

Rabaey K., Angenent L., Schröder U. and Keller J. (2009). Bioelectrochemical Systems: From Extracellular Electron Transfer to Biotechnological Application. IWA Publishing, London, UK.

Rahman A., Wadhawan T., Khan E., Riffat R., Takács I., De Clippeleir H., Wett B., Jimenez J., Al-Omari A. and Pathak B. (2014). Characterizing and quantifying flocculated and adsorbed chemical oxygen demand fractions in high-rate processes. Proceedings of IWA Specialist Conference, Global Challenges: Sustainable Wastewater Treatment and Resource Recovery, Kathmandu, Nepal.

Rahman A., Yapuwa H., Baserba M. G., Rosso D., Jimenez J. A., Bott C., Al-Omari A., Murthy S., Riffat R. and De Clippeleir H. (2017). Methods for quantification of biosorption in high-rate activated sludge systems. *Biochemical Engineering Journal*, **128**, 33–44.

Reijnders L. (2014). Phosphorus resources, their depletion and conservation, a review. *Resources, Conservation and Recycling*, **93**, 32–49.

Rittmann B. E. and McCarty P. L. (2001). Environmental Biotechnology: Principles and Applications. McGraw-Hill, Boston.

Robles Martínez Á. (2013). Modelling, Simulation and Control of the Filtration Process in a Submerged Anaerobic Membrane Bioreactor Treating Urban Wastewater. Editorial Universitat Politècnica de València, Spain.

Rozendal R. A., Hamelers H. V. M., Rabaey K., Keller J. and Buisman C. J. N. (2008). Towards practical implementation of bioelectrochemical wastewater treatment. *Trends in Biotechnology*, **26**(8), 450–459.

Sancho I., Lopez-Palau S., Arespacochaga N. and Cortina J. L. (2019). New concepts on carbon redirection in wastewater treatment plants: A review. *Science of The Total Environment*, **647**, 1373–1384.

Sato T., Qadir M., Yamamoto S., Endo T. and Zahoor A. (2013). Global, regional, and country level need for data on wastewater generation, treatment, and use. *Agricultural Water Management*, **130**, 1–13.

Scherson Y. D., Wells G. F., Woo S.-G., Lee J., Park J., Cantwell B. J. and Criddle C. S. (2013). Nitrogen removal with energy recovery through N_2O decomposition. *Energy & Environmental Science*, **6**(1), 241–248.

Scherson Y. D., Woo S.-G. and Criddle C. S. (2014). Production of nitrous oxide from anaerobic digester centrate and its use as a co-oxidant of biogas to enhance energy recovery. *Environmental Science & Technology*, **48**(10), 5612–5619.

Schreiber F., Wunderlin P., Udert K. M. and Wells G. F. (2012). Nitric oxide and nitrous oxide turnover in natural and engineered microbial communities: biological pathways, chemical reactions, and novel technologies. *Frontiers in Microbiology*, **3**, 372.

Service R. F. (2018). Ammonia – a renewable fuel made from sun, air, and water – could power the globe without carbon. https://www.sciencemag.org/news/2018/07/ammonia-renewable-fuel-made-sun-air-and-water-could-power-globe-without-carbon/ (accessed 08 September 2019).

Shapleigh J. P. (2013). Denitrifying prokaryotes. In: The Prokaryotes, E. Rosenberg, E. F. DeLong, S. Lory, E. Stackebrandt and F. Thompson (eds), Springer, Berlin, Heidelberg, pp. 405–425.

Smith K., Liu S., Liu Y. and Guo S. (2018). Can China reduce energy for water? A review of energy for urban water supply and wastewater treatment and suggestions for change. *Renewable and Sustainable Energy Reviews*, **91**, 41–58.

Song Y., Xiao L., Jayamani I., He Z. and Cupples A. M. (2015). A novel method to characterize bacterial communities affected by carbon source and electricity generation in microbial fuel cells using stable isotope probing and Illumina sequencing. *Journal of Microbiological Methods*, **108**, 4–11.

Tao T. and Xin K. L. (2014). Public health: A sustainable plan for China's drinking water. *Nature News*, **511**(7511), 527.

Tchobanoglous G., Burton F. L. and Stensel H. D. (2003). Wastewater Engineering Treatment and Reuse. McGraw-Hill Higher Education, Boston, US.

Theis T. and Hicks A. (2012). Methanol Use in Wastewater Denitrification. Exponent, Alexandria, USA.

Van Winckel T. (2014). Development of High-Rate Activated Sludge Processes for Energy-Efficient Wastewater Treatment, *MSc. thesis*, Gent University, Belgium.

Vincent L., Michel L., Catherine C. and Pauline R. (2014). The energy cost of water independence: the case of Singapore. *Water Science and Technology*, **70**(5), 787–794.

Wan J., Gu J., Zhao Q. and Liu Y. (2016). COD capture: a feasible option towards energy self-sufficient domestic wastewater treatment. *Scientific Reports*, **6**, 25054.

Wang R., Wilfert P., Dugulan I., Goubitz K., Korving L., Witkamp G.-J. and van Loosdrecht M. C. (2019). Fe (III) reduction and vivianite formation in activated sludge. *Separation and Purification Technology*, **220**, 126–135.

Wett B., Buchauer K. and Fimml C. (2007). Energy self-sufficiency as a feasible concept for wastewater treatment systems. IWA Leading Edge Technology Conference. Asian Water, Singapore, pp. 21–24.

Wilfert P., Mandalidis A., Dugulan A., Goubitz K., Korving L., Temmink H., Witkamp G. and Van Loosdrecht M. (2016). Vivianite as an important iron phosphate precipitate in sewage treatment plants. *Water Research*, **104**, 449–460.

Xu G. R., Yan Z. C., Wang Y. C. and Wang N. (2009). Recycle of Alum recovered from water treatment sludge in chemically enhanced primary treatment. *Journal of Hazardous Materials*, **161**(2), 663–669.

Xu L., Dong F., Zhuang H., He W., Ni M., Feng S.-P. and Lee P.-H. (2017). Energy upcycle in anaerobic treatment: Ammonium, methane, and carbon dioxide reformation through a hybrid electrodeionization–solid oxide fuel cell system. *Energy Conversion and Management*, **140**, 157–166.

Yin Y., Liu Y.-J., Meng S.-J., Kiran E. U. and Liu Y. (2016). Enzymatic pretreatment of activated sludge, food waste and their mixture for enhanced bioenergy recovery and waste volume reduction via anaerobic digestion. *Applied Energy*, **179**, 1131–1137.

Yu J., Seon J., Park Y., Cho S. and Lee T. (2012). Electricity generation and microbial community in a submerged-exchangeable microbial fuel cell system for low-strength domestic wastewater treatment. *Bioresource Technology*, **117**, 172–179.

Zhang F., Ge Z., Grimaud J., Hurst J. and He Z. (2013). Long-term performance of liter-scale microbial fuel cells treating primary effluent installed in a municipal wastewater treatment facility. *Environmental Science & Technology*, **47**(9), 4941–4948.

Zhang M., Gu J. and Liu Y. (2019). Engineering feasibility, economic viability and environmental sustainability of energy recovery from nitrous oxide in biological wastewater treatment plant. *Bioresource Technology*, **282**, 514–519.

Zhao W., Ting Y., Chen J., Xing C. and Shi S. (2000). Advanced primary treatment of waste water using a bio-flocculation-adsorption sedimentation process. *Acta Biotechnologica*, **20**(1), 53–64.

Zou S. and He Z. (2018). Efficiently "pumping out" value-added resources from wastewater by bioelectrochemical systems: A review from energy perspectives. *Water Research*, **131**, 62–73.

Zou S., Qin M., Moreau Y. and He Z. (2017). Nutrient-energy-water recovery from synthetic sidestream centrate using a microbial electrolysis cell – forward osmosis hybrid system. *Journal of Cleaner Production*, **154**, 16–25.

Chapter 3

Integrated A-B processes for municipal wastewater treatment

Jun Gu and Yu Liu

3.1 INTRODUCTION

Currently, the widely adopted conventional biological processes for municipal wastewater treatment are receiving increasing attention regarding their intensive energy consumption and huge amount of sludge production. However, significant improvements in energy efficiency are unlikely to be reached via further optimization of the conventional biological processes (Wan *et al.*, 2016). Therefore, more effort should be dedicated to the development and adoption of innovative process configurations and emerging technologies for municipal wastewater treatment. In fact, chemical oxygen demand (COD) capture from municipal wastewater is critical to move towards direct energy recovery instead of biological oxidation of COD. In Chapter 2, novel technologies that are able to capture COD for energy recovery have been presented in detail. However, these technologies could not stand alone for municipal wastewater treatment due to their ineffectiveness in nitrogen removal. As such, processes for COD capture should be integrated with processes specially designed for energy-efficient nitrogen removal, and it is this configuration which has become known as the A-B process. In the A-B process, the A-stage is designed for the capture of COD for energy recovery instead of biological oxidation, whereas the B-stage is primarily dedicated to nutrient removal or recovery (Wan *et al.*, 2016). As shown in Figure 3.1, there seem to be various combinations of the A-B process.

Author(s): Yu Liu, Jun Gu and Meng Zhang
doi: 10.2166/9781789060089_0069

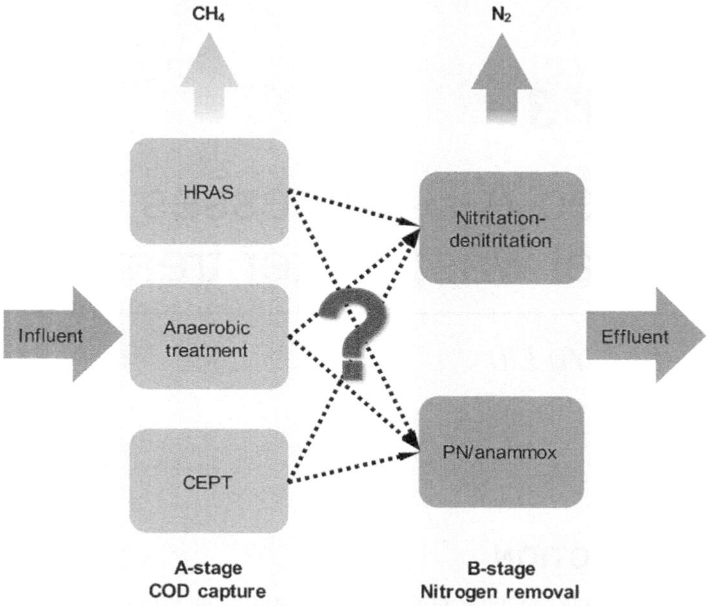

Figure 3.1 Potential combinations of A-B processes for energy-efficient municipal wastewater treatment. HRAS: High rate activated sludge; CEPT: Chemically enhanced primary treatment; PN: Partial nitrification. Adapted from Wan *et al.* (2016).

However, to achieve organic integration of the A- and B-stages, the effectiveness of COD capture at the A-stage and nitrogen removal at the B-stage should be considered. More specifically, the selection of energy-effective nitrogen removal processes for the B-stage largely depends on the characteristics of the A-stage effluent, e.g. the COD/N ratio which is mainly governed by the extent of carbon capture at A-stage. Moreover, it is necessary to evaluate the engineering feasibility and economic viability of integrated A-B processes in a holistic manner with the aim of providing technological solutions for a more energy-efficient reclamation of municipal wastewater.

Therefore, the objectives of this chapter are: (i) to present various combinations of A-B processes; (ii) to look into the performances of these A-B processes through mass flow and balance; (iii) to evaluate the potential energy recovery and energy consumption in the respective A-B processes; and (iv) to pinpoint the potential challenges associated with the integrated A-B processes. An overall energy gain in wastewater treatment plants (WWTPs) appears to be very achievable with the A-B process as a core.

3.2 INTEGRATED MICRO-AERATION AND NITRITATION-DENITRITATION PROCESS

3.2.1 Process performance

This A-B process consists of a moving bed biofilm reactor (MBBR) as the A-stage followed by a sequencing batch reactor (SBR) with step-feed mode as the B-stage for treating municipal wastewater (Xu *et al.*, 2017). The MBBR packed with 30% of gel beads was operated with a dissolved oxygen (DO) concentration below 0.2 mg/L through micro-aeration, and a short hydraulic retention time (HRT) of 2.6 hours. The B-stage was a step-feed SBR operated with 4 alternating feed-anoxic-oxic sub-cycles. The DO concentrations in the oxic phases were maintained in the range of 1.4–1.6 mg/L, while the HRT and sludge retention time (SRT) were controlled at 5.5 hours and 4–5 days, respectively. The experiments were carried out in phases as described in Table 3.1.

In the A-stage micro-aerated MBBR, the influent COD and NH_4^+-N were reduced by about 55% and 15%, respectively, at a DO concentration below 0.2 mg/L. The effluent COD mainly contained 0.91 mM of acetic acid and 0.41 mM of propionic acid, corresponding to a COD concentration of about 100 mg/L, with an average sludge concentration of 65.3 mg Volatile Suspended Solids (VSS)/L in the effluent. In the B-stage step-feed SBR, more than 95% of ammonium was removed through nitritation-denitritation in period I, while lower phosphorus removal efficiency was observed, probably due to the low influent COD/N ratio of 3.2 to the B-stage. To achieve concurrent nitrogen and phosphorus removal, the influent COD/N ratio was further increased from 7 to 8 in period II, resulting in a higher COD/N ratio of 4.5 in the influent to the B-stage. Although the nitrite accumulation rate (NAR) slightly declined from 76% during period I to 72% during periods II-IV, the ammonium and total nitrogen (TN) removal efficiencies rose to 99% and 88%, respectively. Meanwhile, the removal efficiency of phosphorus was increased to 89% in period II and further to 93% in period III, but it was observed that the further rise of the influent phosphorus concentration in period IV significantly deteriorated biological phosphate removal.

Table 3.1 Operating conditions of the integrated micro-aeration and nitritation-denitritation process. (Xu *et al.*, 2017, with permission).

Period	Time, d	COD, mg/L	NH_4^+-N, mg/L	PO_4^{3-}-P, mg/L	COD/N
I	1–73	334.0 ± 67.2	47.4 ± 5.3	4.5 ± 1.5	7.0
II	74–103	335.9 ± 59.3	41.8 ± 5.4	4.5 ± 1.5	8.0
III	104–137	325.1 ± 65.8	40.8 ± 3.4	7.6 ± 0.9	8.0
IV	137–199	305.7 ± 46.5	38.2 ± 3.5	12.0 ± 1.3	8.0

3.2.2 Nutrients removal in a B-stage step-feed SBR

In the process described in Figure 3.2, the ammonium was firstly oxidized to nitrite in the oxic sub-cycles, which was then denitrified in the subsequent anoxic sub-cycles. The nitrogen profiles in the four oxic sub-phases are shown in Figure 3.3, according to which the specific activities of AOB and NOB could be determined. Throughout the four oxic sub-cycles, the average specific AOB activity was found to be 19 mg NO_2^--N/(g VSS h), while the specific NOB activity averaged at 7.3 mg NO_3^--N/(g VSS h), i.e. the activity of AOB was almost 2.6 times higher than that of NOB. This in turn may explain the high NAR, ranging from 62% to 70%, achieved in the oxic sub-cycles. As can be seen in Figure 3.3, in the first 4-min of each oxic sub-phase, nitrate production appeared to be negligible. This suggested that NOB activity could not be recovered quickly from the preceding anoxic phase. The inhibition on NOB activity might be due to the rapid shift of anoxic and oxic conditions, which was crucial for achieving a sustainable nitritation (Gilbert *et al.*, 2014). It was also found that the lag phase for NOB activity recovery was about 4 min, accounting for 33% of the oxic sub-cycle duration, i.e. there was 33% more time for AOB to oxidize ammonium, but with no further nitrite oxidation in all oxic sub-cycles. Furthermore, a short SRT under relatively high temperature has been proven to be effective for washing out NOB from a system (Hellinga *et al.*, 1998). As such, the SRT in the step-feed SBR (Figure 3.2) was controlled at 4–5 days under 30°C, favouring suppression of NOB.

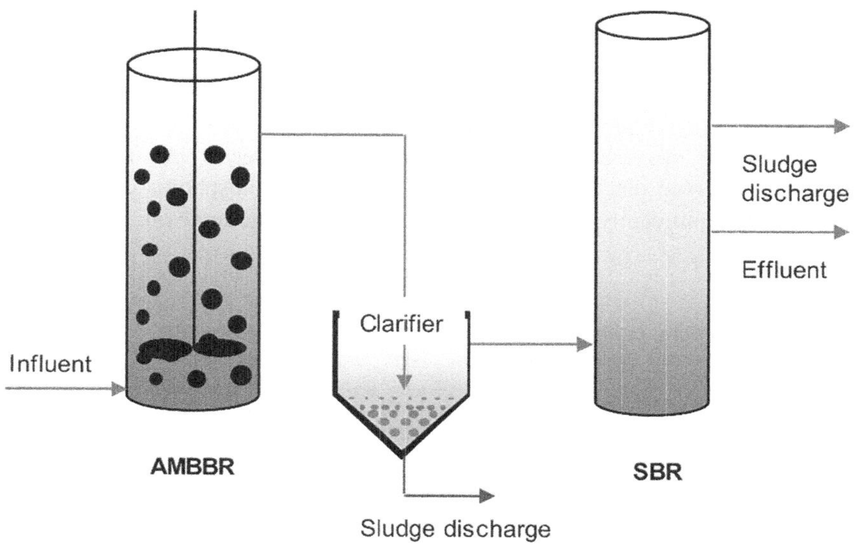

Figure 3.2 An A-B process with a micro-aerated MBBR as A-stage and nitritation-denitritation as B-stage. Adapted from Xu *et al.* (2017), with permission.

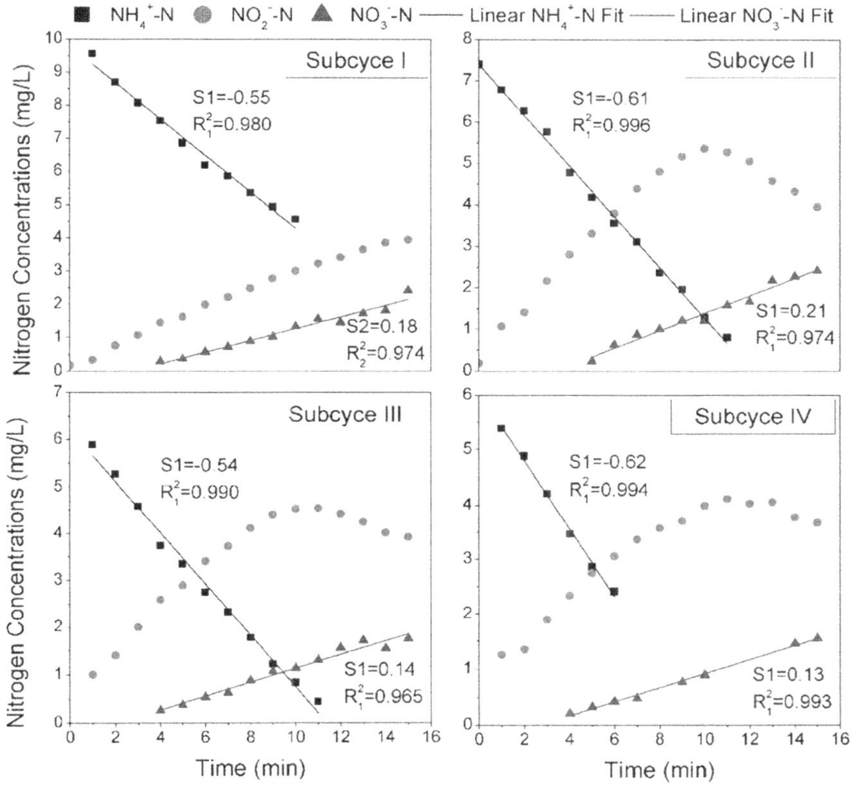

Figure 3.3 Cyclic nitrogen profiles in the aerobic sub-cycles of the step-feed SBR. Xu *et al.* (2017), with permission.

In period IV, ammonium was converted to nitrite with negligible nitrate generation in all the oxic sub-phases, and the average production rate of nitrite (i.e. 17.6 mg NO_2^--N/(g VSS h)) was determined to be 44 times higher than that of nitrate (i.e. 0.4 mg NO_3^--N/(g VSS h)). These results obviously suggested that a successful partial nitrification was achievable in this system. In addition, the produced nitrite in the oxic sub-phases was quickly denitrified to nitrogen gas in the subsequent anoxic sub-phases with VFA produced from the A-stage as a carbon source. This in turn created a situation where the availability of nitrite for NOB was limited at the beginning of the oxic sub-phases, evidenced by 80–88% of nitrite removal by the end of the anoxic sub-phases.

Excellent phosphorus removal was also observed in the step-feed SBR where phosphorous release and uptake were clearly observed in the anoxic and oxic sub-phases. The specific phosphorus release rate, in anoxic sub-cycles A2–A4, averaged at 15.0 mg PO_4^{3-}-P/(g VSS h), while the release rate observed in A1 reached 22.9 mg PO_4^{3-}-P/(g VSS h). The high phosphorus release rate in A1 was likely due to the fact that higher COD concentration was available for promoting phosphorus release in the anoxic sub-cycles, where the amount of nitrite was negligible at the beginning of A1. In fact, it is known that higher COD concentration favours phosphorus release. By contrast, the phosphorus uptake rates were found to be higher than 33.6 mg PO_4^{3-}-P/(g VSS h) in each oxic chamber. It has been revealed that nitrite with a concentration higher than 8 mg N/L could adversely affect anoxic and aerobic phosphorus uptake (Meinhold *et al.*, 1999). In the work by Xu *et al.* (2017), nitrite concentration was found to be lower than the inhibition threshold over the whole operating period. Thus, the inhibitory effect from nitrite on phosphorus removal might be insignificant for the B-stage SBR. A similar study was also reported in a continuous pilot step-feed MBR treating primarily settled municipal wastewater. As reported by Wang *et al.* (2019), a single parameter-based operation strategy could not lead to effective repression of NOB, instead DO and SRT should be synergistically controlled. In addition to DO and SRT, the sludge return ratio also played a vital role in establishing stable nitrite shunt. In fact, this caused the rapid shift of anoxic/oxic conditions in the step-feed SBR (Figure 3.2), e.g. the observed HRT in each oxic chamber was not long enough for NOB recovery from the anoxic disturbances when the sludge return ratio was increased to 2.5 (Wang *et al.*, 2019).

3.2.3 COD flows in a micro-aeration and nitritation-denitritation process

Mass balance analysis of COD was performed using the data obtained in period III of the study as described in Table 3.1. It was revealed that 55% of the influent COD was removed in the A-stage MBBR, while the remaining 45% entered the B-stage for nutrient removal. The suspended sludge in the effluent of the A-stage MBBR was as low as 91 mg COD/L, equivalent to 28% of the influent COD if the empirical formula of $C_5H_7O_2N$ was assumed for biomass. Similar to COD oxidation in the CAS process, 55% of the influent COD was removed in the A-stage micro-aerated MBBR at a DO concentration lower than 0.2 mg/L. At the B-stage, the effluent COD and the excess sludge accounted for 5% and 31% of the influent COD, respectively, implying that 9% of the COD was lost in nitritation-denitritation. In the integrated process (Figure 3.4), about 21% of COD was captured as methane gas through anaerobic digestion (AD) of waste sludge, and this was about 10% lower than that in a typical CAS process (Wan *et al.*, 2016). In fact, such observed differences in biogas production likely resulted from the different biodegradabilities of the primary sludge in the CAS process and the sludge from the micro-aerated MBBR. It appears from the literature that about 65% of primary sludge-COD in the CAS process could be converted to methane due to the high content of readily biodegradable organics (Parkin &

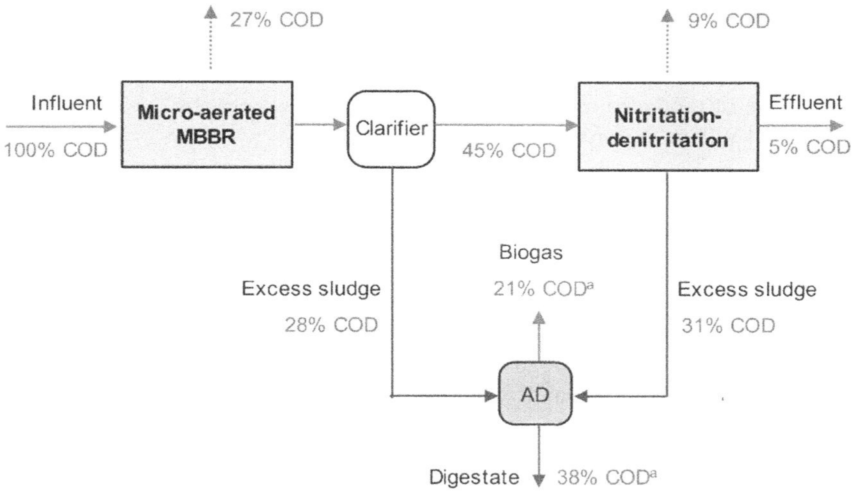

Figure 3.4 COD flows in the integrated micro-aeration and nitritation-denitritation process. [a]Data calculated according to Parkin and Owen (1986).

Owen, 1986), whereas only 30–50% could be converted for the waste sludge from an MBBR due to the relatively low sludge destruction rate (Cao & Pawłowski, 2012; Nges & Liu, 2010). In fact, in the process configuration presented in Figure 3.2, the A-stage micro-aerated MBBR possessed two major functions, i.e. (i) to reduce influent COD concentration to a level sufficient for simultaneous nitrogen and phosphorus removal at low DO concentration; and (ii) to convert complex organics to VFAs which are known to be beneficial for fast denitrification and phosphorus uptake (Lopez-Vazquez *et al.*, 2009; Guerrero *et al.*, 2011).

3.2.4 Cost-benefit analysis

Sludge production in the micro-aeration and nitritation-denitritation process averaged at 135.3 mg VSS/L, corresponding to a sludge yield of 0.44 g VSS/g COD removed, which was comparable with that obtained in the CAS process (e.g. 0.3–0.5 g dry biomass/g COD removed). It has been reported that aeration-associated energy consumption accounts for almost 50% of in-plant energy consumption in WWTPs with CAS process as a core (Henderson, 2002). In general, in current WWTPs, a substantial amount of energy is required to drive nitrification. For example, in one study, implementation of nitrification increased the energy consumption for aeration from 0.305 kWh/m^3 to 0.405 kWh/m^3 (Monteith *et al.*, 2007). Compared to full nitrification-denitrification, energy consumption due to aeration could be decreased by 25% through nitritation-denitritation at the B-stage. Meanwhile, additional energy saving could also be achievable at the micro-aerated A-stage for limited COD removal.

According to the COD mass balance (Figure 3.4), 21% of the influent COD was eventually converted to methane gas through AD of excess sludge, equivalent to 0.024 m^3 methane/m^3 wastewater treated. The electricity generated via combustion of produced methane, $E_{methane}$ (kWh/m^3 wastewater), was calculated to be 0.09 kWh/m^3 according to Equation 3.1:

$$E_{methane} = Q_{methane} \times CV_{methane} \times f \times e \tag{3.1}$$

where $Q_{methane}$ is the production rate of methane in m^3 methane/m^3 wastewater treated; $CV_{methane}$ is the methane combustion value, i.e. 40 MJ/m^3 methane (Gupta *et al.*, 2015); f is the engine efficiency for electricity production, i.e. 0.35 (Cao, 2011); e is the coefficient for energy unit conversion from g MJ to kWh, at a typical value of 0.28. Although this process (Figure 3.2) showed a lower energy recovery potential as compared with the CAS process equipped with AD, the aeration-associated energy consumption could be cut by about 50% for COD removal and 25% for nitrogen removal. Overall, the A-B process presented in Figure 3.2 could provide a cost-effective alternative for improved biological nutrients removal in municipal wastewater treatment with reduced energy consumption.

3.3 INTEGRATED CEPT AND NITRITATION-DENITRITATION PROCESS

3.3.1 Process configuration

Chemically enhanced primary treatment (CEPT) has been employed to treat municipal wastewater for decades. Basically, it is a physical-chemical treatment of wastewater by the addition of chemical coagulants or flocculants into wastewater to enhance the total suspended solids (TSS) removal during primary treatment. Figure 3.5 shows the integrated CEPT and nitritation-denitritation process with AD. Eighty percent of TSS in wastewater could be removed by CEPT, together with 60% removal of particulate COD in full-scale WWTPs (Metcalf & Eddy, 2014; Parker *et al.*, 2001; Wang *et al.*, 2009). Compared to conventional primary clarification, the amount of sludge generated from the CEPT process was found to increase by about 45%, of which 33% resulted from improved solids capture. However, CEPT is ineffective in capturing soluble COD from municipal wastewater, which accounts for more than 30% of total COD (Rossle & Pretorius, 2001; Xu *et al.*, 2006). Overall, 60% of total COD could be removed through CEPT (Metcalf & Eddy, 2014; Morrissey & Harleman, 1992).

Given COD and ammonium concentrations of 500 mg/L and 40 mg NH_4^+-N/L, respectively, in a typical municipal wastewater (McCarty *et al.*, 2011), the COD/N ratio after CEPT can be reduced to about 5. This implies that in processes where CEPT serves as the A-stage, the soluble COD entering the B-stage might not be sufficient to sustain full denitrification. In this situation, nitritation-denitritation

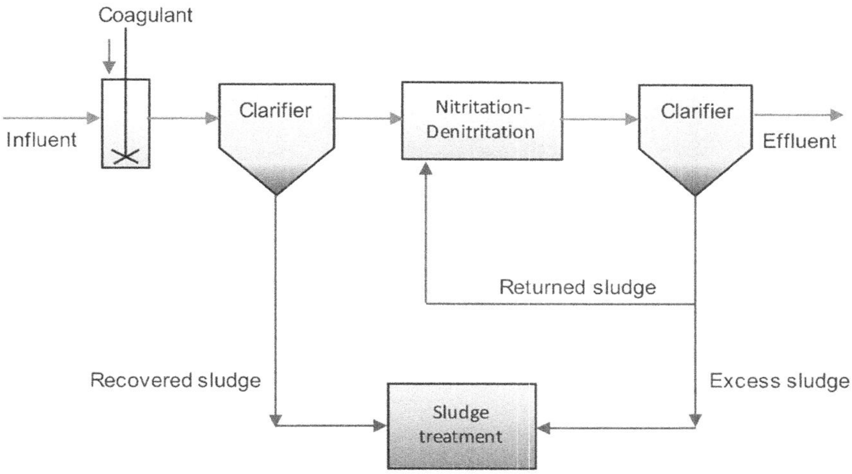

Figure 3.5 The integrated CEPT and nitritation-denitritation process. Adapted from Wan *et al.* (2016).

may offer a feasible solution for the B-stage. The solids captured via CEPT can be directed to AD for energy production, while evidence suggests that adding chemical coagulants in CEPT would not adversely influence the AD for wasted sludge treatment.

3.3.2 COD flows in an integrated CEPT and nitritation-denitritation process

As discussed above, an integrated CEPT and nitritation-denitritation process represents a feasible configuration of the A-B process. As shown in Figure 3.6, the COD mass flows suggested that about 60% of influent COD could be captured by CEPT, with the remaining 40% entering the B-stage for nitrogen removal via nitritation-denitritation (Wan *et al.*, 2016). The excess sludge produced at the B-stage accounts for 12% of influent COD. As a result, about 72% of the influent COD could be directed to an anaerobic digester for energy production and sludge volume reduction. In fact, the sludge captured by CEPT has been found to be rich in organics and more suitable for biogas production through AD. In Figure 3.6, it can be seen that 43% of influent COD could be ultimately recovered as methane, which is comparable with the 37% reported by Rahman *et al.* (2019).

3.3.3 Cost-benefit analysis

According to the COD mass balance presented in Figure 3.6, about 215 mg COD/L would be channeled into methane, equivalent to 0.075 m^3 methane/m^3 wastewater

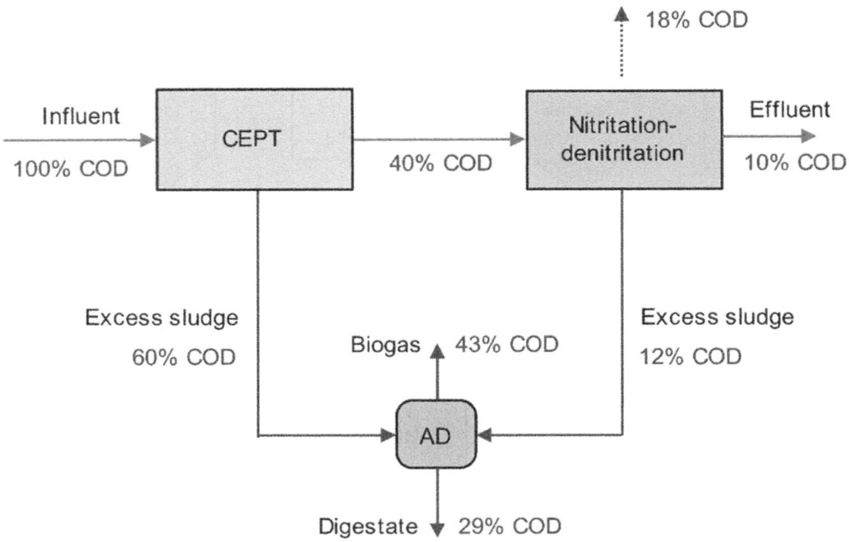

Figure 3.6 COD flows in the integrated CEPT and nitritation-denitritation process. Data from Wan *et al.* (2016).

for municipal wastewater with 500 mg COD/L. As such, the maximum recoverable electrical energy was estimated as 0.25 kWh/m^3 wastewater treated, which could offset more than half of the total energy consumption in WWTPs with the CAS process as a core unit, i.e. 0.47 kWh/m^3 (Drechsel *et al.*, 2015; Zhou *et al.*, 2013).

Despite the high energy recovery potential in the process presented in Figure 3.5, adding chemical coagulants during CEPT inevitably results in an increased operational cost. The chemical cost associated with alum and modified PACl has been reported to be US\$ 0.0186/m^3 wastewater, which induced an additional annual cost of US\$ 4.5 million in the Bailonggang WWTP in Shanghai with a treatment capacity of 1,200,000 m^3/day (Wang *et al.*, 2009). It has also been reported, in a WWTP in Mexico, that an annual operational cost of US\$ 4 million was incurred due to CEPT with a mean treatment capacity of 4.49×10^5 m^3/day, equivalent to US\$ 0.024/m^3 wastewater (Harleman *et al.*, 2002). In addition, the cost associated with CEPT was estimated to be about 0.1 €/m^3 in a wastewater treatment facility with a capacity of 500 m^3/d where both alum and polymer were used (Diamantis *et al.*, 2013). However, it should be noted that an increased amount of sludge is generated during CEPT, e.g. 1.5 to 2 times more sludge produced due to the usage of chemicals (Harleman & Morrissey, 1990; Semerjian & Ayoub, 2003). A study by Morrissey & Harleman (1992) showed that sludge production was increased by 45% after the implementation of CEPT, of which 33% was attributed to enhanced solids capture and 12% from applied chemicals. Obviously, despite the advantages of CEPT in COD capture, the usage of chemicals inevitably results in increased sludge production and high operational costs.

3.4 INTEGRATED HRAS AND NITRITATION-DENITRITATION PROCESS

3.4.1 Process configuration

The high-rate activated sludge (HRAS) process is commonly operated with a short sludge retention time and a short hydraulic retention time (Wett *et al.*, 2007). Recently, the performance of the HRAS process has been improved by incorporating feast–famine cycles, resulting in a total COD reduction of 45–84% (Meerburg *et al.*, 2015). As such, the wastewater COD/N ratio could be lowered to 0.67–8.0 after HRAS treatment (Liu *et al.*, 2018). It is reasonable to consider that this wide range of the COD/N ratio in the HRAS process allows different biological pathways for nitrogen removal in any subsequent B-stage, e.g. nitritation-denitritation, partial nitritation and anammox, and partial denitrification and anammox. This section only focuses on the integrated HRAS and nitritation-denitritation process.

In this integrated process (Figure 3.7), raw wastewater is firstly treated in an HRAS tank for COD capture by returned activated sludge from the clarifier, after which the mixed liquor is further directed to a settling tank for solid and liquid separation. Subsequently, the supernatant is channeled to the nitritation-denitritation unit for nitrogen removal, while the settled sludge is partially returned to the HRAS tank and the rest is subjected to AD for energy recovery.

Among the energy-efficient processes for biological nitrogen removal, nitritation-denitritation has been considered as a more practical option with high

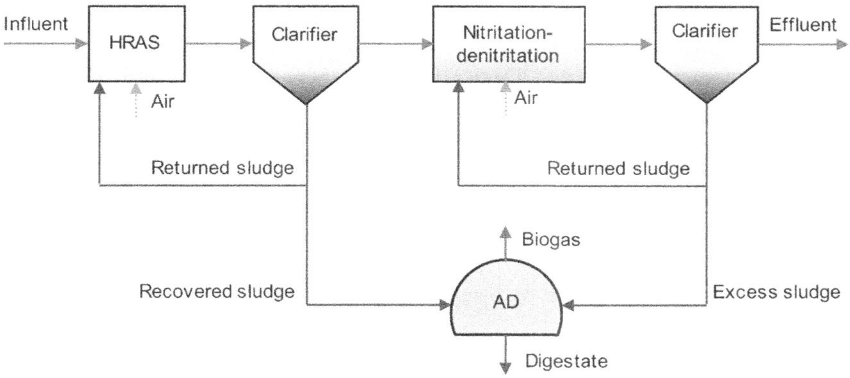

Figure 3.7 Illustration of an integrated HRAS and nitritation-denitritation process. Adapted from Wan *et al.* (2016).

feasibility and stability. Theoretically, the aeration-associated energy for nitritation could be reduced by 25%, together with 40% reduction in COD demand for denitritation (van Kempen *et al.*, 2001). In addition, reduced sludge production is expected during nitritation-denitritation against full nitrification and denitrification.

3.4.2 COD/N ratio in HRAS effluent

A favourable COD/N ratio in the effluent from the HRAS process is crucial for subsequent nitritation-denitritation. Theoretically, denitrifying 1 gram of NO_2^--N requires 1.71 gram of soluble COD (Metcalf & Eddy, 2014), indicating that a minimum COD/N ratio of 1.71 is essential for complete nitrogen removal via nitritation and denitritation. It should be realized that the COD/N ratio is smaller than its stoichiometric value due to the fact that part of the organic carbon is channeled into biomass and carbon dioxide through biological oxidation. In practice, the COD/N ratios required for nitritation-denitritation have been reported to be in the range of 2.0 to 5.0 (Ge *et al.*, 2014; Gu *et al.*, 2018b; Wang *et al.*, 2019). However, a COD/N ratio higher than 7.0 was also proposed for sustainable nitritation-denitritation (Lemaire *et al.*, 2008; Nogaj *et al.*, 2014; Regmi *et al.*, 2014). These discrepancies were likely caused by different wastewater compositions in terms of types of soluble COD and total COD. In fact, it is well known that COD can be classified into particulate COD (pCOD) and soluble COD (sCOD), between which pCOD is not readily biodegradable prior to hydrolysis. However, in municipal wastewater, pCOD indeed represents a non-negligible portion in total wastewater COD, while part of it could be converted to soluble COD during biological processing. This in turn suggests that COD/N ratio might not be a precise parameter for evaluating COD sufficiency for nitritation and denitritation. Instead, readily biodegradable soluble COD is recommended for this purpose. For example, in a pilot HRAS and Modified Ludzack-Ettinger (MLE) process it was found that a COD/N ratio of 6 was not sufficient for denitrifying the produced NO_X^-, resulting in a low nitrogen removal efficiency of 25–63% (Miller *et al.*, 2012). The COD characterization further revealed that sCOD and pCOD in HRAS effluent were 95 and 115 mg/L, respectively, giving an actual sCOD/N ratio of 2.7 only. Compared with HRAS, primary settling has proved effective for pCOD removal without affecting sCOD. As such, in the above case, the primary effluent was blended with the HRAS effluent at a ratio of 1 to 1 to generate a tCOD/N ratio of 7.6 and sCOD/N ratio of 4, leading to a nitrogen removal efficiency of 71–87% (Miller *et al.*, 2012).

3.4.3 Control of COD/N ratio in the HRAS process

In the integrated HRAS and nitritation-denitritation process, a proper COD/N ratio is essential for complete nitritation-denitritation. Excessive COD capture in the HRAS stage lowers the COD/N ratio, and is detrimental to subsequent denitritation (Gu *et al.*, 2017; Miller *et al.*, 2012). It is obvious that COD capture by the HRAS process should not compromise the subsequent denitritation, i.e. a balance between energy recovery and COD reserved for complete denitritation should be maintained during long-term operation. For a typical municipal wastewater with a COD/N ratio of above 10 (Metcalf & Eddy, 2014), the energy recovery through COD capture by HRAS is feasible, but should be well controlled within a certain extent.

To maintain a suitable COD/N ratio in the integrated HRAS and nitritation-denitritation process, HRT, SRT, DO and F/M ratio have been identified as the key parameters to optimize operation, as described in Chapter 2. Currently, extensive efforts have been dedicated to process optimization with the aim of maximizing energy recovery and COD removal efficiency (Jimenez *et al.*, 2015; Meerburg *et al.*, 2015; Meerburg *et al.*, 2016). However, little attention has been given to how to achieve a suitable COD/N for nitritation-denitritation in this integrated process for municipal wastewater treatment, while it definitely deserves further investigation.

Another option to make up the required COD/N ratio is to bypass a portion of primary settled municipal wastewater from HRAS (Figure 3.8). This is based on the understating that primary settling of wastewater does not remove sCOD. As such, the relatively high soluble COD/N ratio becomes achievable in the HRAS effluent. For example, in an integrated process consisting of an HRAS process as the A-stage for COD capture and an anaerobic selector-three aerobic tanks in series – called NiDeMA (Nitritation Denitritation through Modulating Aeration) – as the B-stage for nitritation-denitritation, the feed to the B-stage was blended with 50% primary effluent Miller *et al.* (2012). It should be pointed out that the

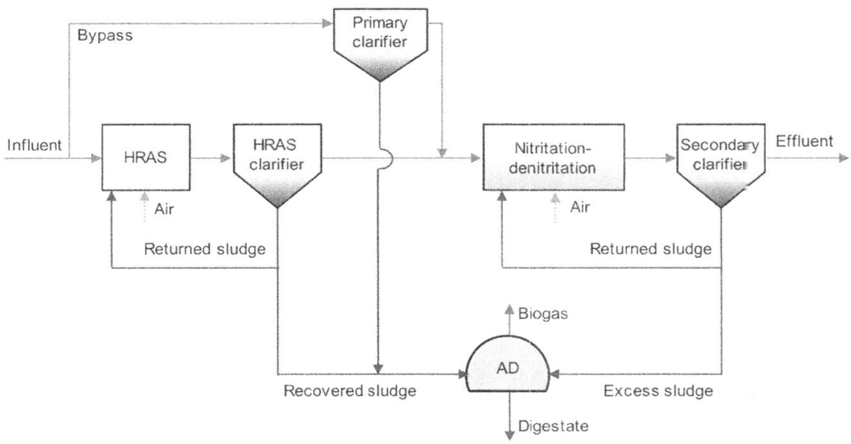

Figure 3.8 A bypass flow-based control strategy for adjusting COD/N ratio in the integrated HRAS and nitritation-denitritation process. Inspired from Miller *et al.* (2012).

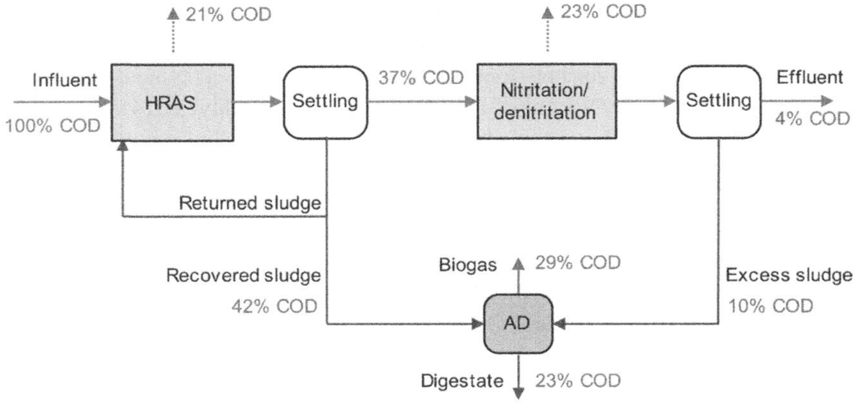

Figure 3.9 COD flows in an integrated HRAS and nitritation-denitritation process. Data from Rahman *et al.* (2019).

bypass ratio depends on the incoming wastewater characteristics, performances of the HRAS unit and primary clarifier.

3.4.4 COD balance in integrated HRAS and nitritation-denitritation process

The COD balance in an integrated HRAS and nitritation-denitritation process is presented in Figure 3.9 using data from Rahman *et al.* (2019), which is derived from a pilot system treating wastewater with 713 mg COD/L and 48.5 mg N/L. It can be seen that about 21% of the influent COD was removed through bio-oxidation in the HRAS unit, while 42% was captured for AD. Meanwhile, the effluent from the HRAS unit with a COD/N ratio of 5.4 was fed to the B-stage nitritation-denitritation unit. As a minimum COD/N ratio of 1.71 is required for denitritation, theoretically the COD concentration in the HRAS effluent should be sufficient for complete denitritation at the B-stage. At the B-stage, about 4% of influent COD was found in the effluent, while another 10% of influent COD was channeled into biosludge. These show that a total of 52% of COD was directed to AD for energy production and sludge volume reduction. Consequently, about 29% of the influent was ultimately recovered as methane gas, which was comparable with that achieved in the CAS process (Wan *et al.*, 2016).

3.5 INTEGRATED HRAS AND PARTIAL NITRITATION-ANAMMOX PROCESS

3.5.1 Process configuration

The high COD/N ratio in municipal wastewater is one of the major challenges in large-scale applications of partial nitritation/anammox. At high COD/N ratios,

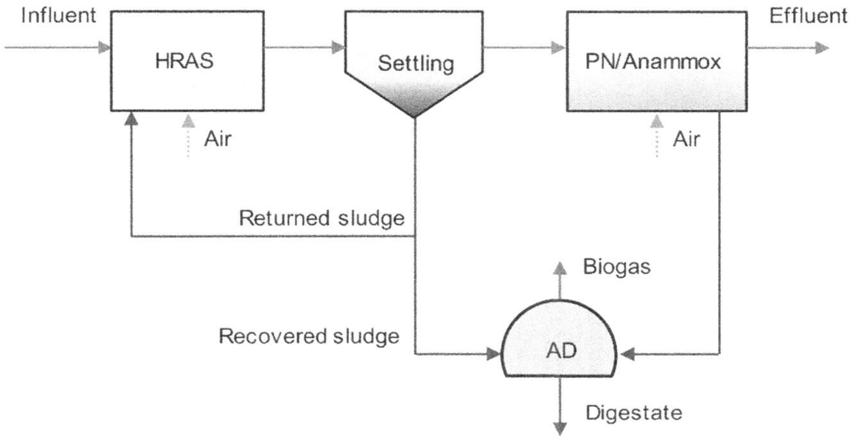

Figure 3.10 Illustration of integrated HRAS and mainstream deammonification. Adapted from Liu *et al.* (2018), with permission.

denitrifiers may outcompete with anammox bacteria for nitrite (Xu *et al.*, 2015). Hence, pretreatment of municipal wastewater is needed to decrease the COD/N ratio through COD capture. As discussed in Section 3.4, the HRAS process has been considered as one of the feasible options for generating a COD/N ratio favourable for mainstream deammonification (Wett *et al.*, 2013).

In this integrated process shown in Figure 3.10, the HRAS unit serves as a pretreatment unit, after which the mixed liquor is directed to a settling tank for solid and liquid separation, while the supernatant is channeled to the B-stage for nitrogen removal through partial nitritation and anammox. Meanwhile, the settled sludge is partially returned to the HRAS tank and the rest is subjected to AD for biogas production. The partial nitritation and anammox process can have either a single-stage or two-stage configuration (Lemaire *et al.*, 2014; Lotti *et al.*, 2015; Regmi *et al.*, 2015).

3.5.2 Impact of COD/N ratio on partial nitritation-anammox

The COD/N ratio plays a vital role in the mainstream deammonification process. As presented above, the COD/N ratio in the HRAS effluent may vary in a wide range of 0.67 to 8 (Table 3.2), probably due to the differences in wastewater characteristics as well as in the operation conditions of HRAS processes. These suggest that the HRAS process may produce an effluent of low COD/N ratio for subsequent mainstream deammonification. However, a study by Miller *et al.* (2015) achieved a relatively high COD/N ratio of 8. As such, the majority of TN was removed through nitritation and denitritation in a two-stage partial nitritation-anammox process, i.e. 56% of TN was removed at the first stage, and anammox seemed to

Table 3.2 Integrated HRAS and partial nitritation/anammox processes for municipal wastewater treatment.

Process at B-Stage	Reactor Type	COD/N	Influent N, mg/L	TN Removal by PN/AMX	NO_3^-/NH_4^+ Conversion Ratio	References
Single-stage PN/AMX	Plug flow	0.67	26.8	33–46%	0.35–0.41	Lotti *et al.* (2015)
Single-stage PN/AMX	MBBR	3	21	73%	0.19	Laureni *et al.* (2016)
Single-stage PN/AMX	IFAS[a]	2.8	40–60	50%	0.15	Lemaire *et al.* (2014)
Single-stage PN/AMX	MBBR	1	49	70–75%[b]	0.1–0.2	Lackner *et al.* (2015)
Two-stage PN/AMX	AVN+MBBR	8	38	13.4%	0.12	Miller *et al.* (2015)

[a]Process maintained by frequent alternating flows between sidestream and mainstream
[b]Denitrification accounted for 14% of TN removal.
AMX = anammox.

play a minor role for nitrogen removal at the second stage, contributing to 13.4% of the TN removal. Obviously, a higher COD/N ratio in the HRAS effluent promotes denitritation activity, while suppressing anammox activity. Moreover, the overall efficiency of energy recovery would also be compromised due to the ineffective COD capture by HRAS. In addition to $12 \pm 5\%$ of COD lost due to bio-oxidation in the HRAS process, another 32% of influent COD was oxidized in the ammonia versus NOx-N (AVN) process, resulting in a total 45% of COD loss through bio-oxidization at the expense of energy consumption (Miller *et al.*, 2015). It should also be noted that the ineffective COD capture by HRAS also promotes sludge production. In the integrated process presented in Figure 3.10, 47% of influent COD ended up in sludge, of which 70% was produced in the HRAS process and 30% in the AVN process. Compared with secondary waste activated sludge (WAS), the sludge generated in this process had a much higher volatile solids reduction (VSR) rate of 82% against 45% for WAS. In addition, the specific methane yield of the sludge from the HRAS process reached above 1.0 m^3 CH_4/kg COD removed, which was also superior to that of WAS, e.g. generally below 0.75 m^3 CH_4/kg COD removed (Miller *et al.*, 2015).

It is believed that the effective repression of NOB is one of the biggest challenges in the mainstream deammonification process. As summarized in Table 3.2, relatively low nitrogen removal efficiency had been observed in the integrated HRAS and partial nitritation-anammox process, likely due to ineffective NOB repression.

3.5.3 COD flows in integrated HRAS and partial nitritation-anammox process

Figure 3.11 shows that 32% of influent COD was harvested at the A-stage (i.e. HRAS) and subjected to AD for biogas production, while 56% remained in the A-stage effluent; 34% of the COD entering the B-stage was oxidized, while 15% was channeled into biosludge. Due to the high VSR of sludge harvested from the HRAS unit, the recoverable methane gas was increased to 33% of influent COD in the B-stage, which was comparable with the performance of the integrated HRAS and nitritation-denitritation process (see Section 3.4.4 above), but lower than that in the integrated CEPT and nitritation-denitritation process (Section 3.3.2 above). A similar observation has also been reported in the Strass WWTP, with a methane gas production of 35.4% (Wett *et al.*, 2007).

3.5.4 Challenges in the integrated HRAS and partial nitritation-anammox process

In general, HRAS effluent possesses a relatively high COD/N ratio which favours the growth of denitrifiers against anammox bacteria in the subsequent deammonification process, causing process instability during long-term operation. Although a prolonged SRT is desirable to achieve a low COD/N ratio (Meerburg

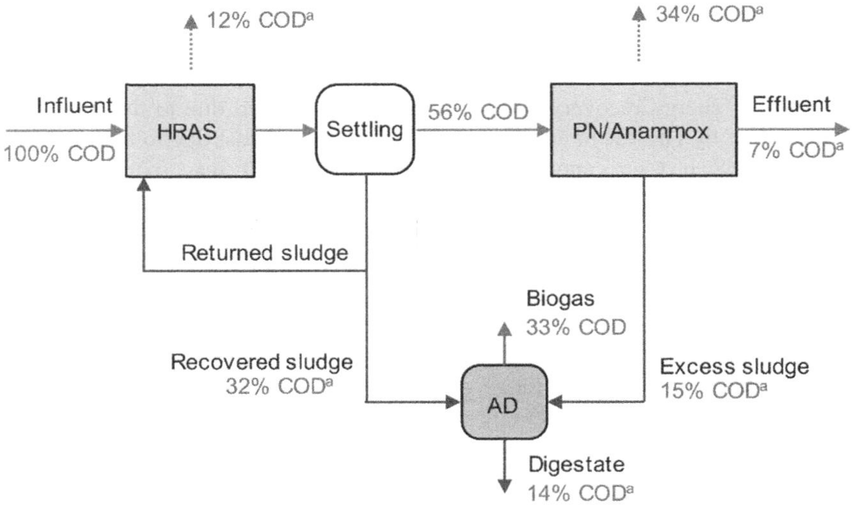

Figure 3.11 COD flows in the integrated HRAS and PN/anammox process. [a]Data from Miller *et al.* (2015).

et al., 2016; Miller *et al.*, 2015), less sludge would be directed to AD for biogas production. On the other hand, current operation practice shows that balancing the sludge sent for AD and that returned to HRAS for COD capture is a great challenge at a short SRT (Meerburg *et al.*, 2016; Xu *et al.*, 2015). Moreover, the performance of the HRAS process is also related to temperature, e.g. low COD capture efficiency was observed in winter (Smitshuijzen *et al.*, 2016).

3.6 INTEGRATED ANAEROBIC AND NITRITATION-DENITRITATION PROCESS

3.6.1 Process configuration

Various anaerobic processes have been developed for municipal wastewater treatment, in which no aeration is needed, and a significantly reduced sludge production is achievable. This makes anaerobic processes more economically viable and energy efficient. Generally, 60–92% of COD removal has been reported in anaerobic processes for municipal wastewater treatment (Dai *et al.*, 2011; Florencio *et al.*, 2001; Kim *et al.*, 2010), depending on the type of anaerobic reactor, operating conditions, wastewater characteristics, etc. However, it should be realized that anaerobic processes are unable to remove soluble nutrients present in municipal wastewater (e.g. ammonium and phosphate) due to the nature of anaerobic reactions. Given this situation, the nitritation-denitritation process has been considered as a feasible option to serve as the B-stage for

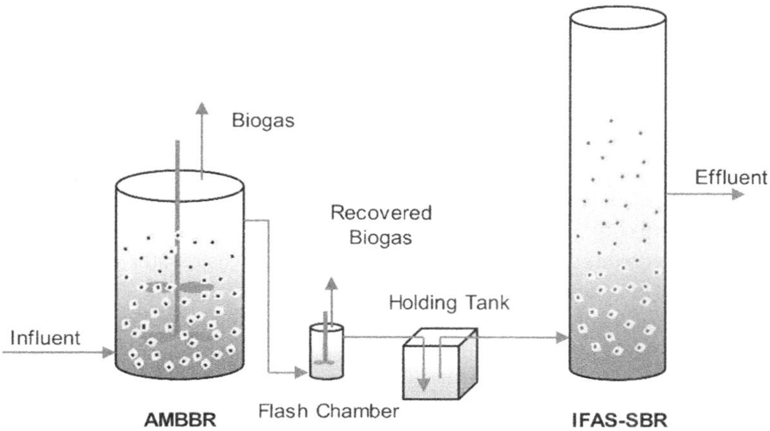

Figure 3.12 Schematic illustration of integrated AMBBR and IFAS-SBR process. Adapted from Gu *et al.* (2017), with permission.

energy-efficient nitrogen removal. Based on this, an integrated anaerobic moving bed biofilm reactor (AMBBR) and integrated fixed-film activated sludge-sequencing batch reactor (IFAS-SBR) process had been developed, in which AMBBR served as the A-stage for COD capture, with IFAS-SBR being the B-stage for biological nitrogen removal through nitritation-denitritation (Gu *et al.*, 2017). Meanwhile, to tackle the issue of dissolved methane, a flash chamber was designed for dissolved methane recovery from the AMBBR effluent (Figure 3.12).

3.6.2 Performance of the integrated AMBBR and IFAS-SBR process

The wastewater fed to the integrated process contained about 400 mg COD/L and 45 mg NH_4^+-N/L (Gu *et al.*, 2017). In the first 20 days of operation, about 68% of COD was removed in the AMBBR, and the COD concentration in the AMBBR effluent was gradually stabilized at a level of 57 mg/L, corresponding to a COD removal efficiency of about 85% at steady state. The biogas production reached 280 L biogas/(m^3 day), of which methane gas accounted for 70%, while 17.1 mg/L of dissolved methane was detected in the AMBBR effluent, which was close to the theoretical value of dissolved methane, i.e. 18.6 mg/L at 30°C (Liu *et al.*, 2014; Yamamoto *et al.*, 1976). In consideration of gaseous and dissolved methane, the total methane yield in the AMBBR was estimated to be 0.23 L/g COD removed, in which dissolved methane accounted for 35%. It should be noted that dissolved methane content is temperature dependent, indicating that a further increase in dissolved methane is expected at lower operational temperature (Smith *et al.*, 2013). Methane has been recognized as a potent greenhouse gas, which has a 25-times higher greenhouse power than carbon dioxide (Forster *et al.*, 2007). Obviously, if dissolved methane produced from anaerobic treatment of wastewater was not properly handled, this will not only affect the overall energy recovery but may also worsen the control of greenhouse gas emissions. Given such a challenging situation, Gu *et al.* (2017) developed a flash chamber for recovering dissolved methane from anaerobic effluent, and results show that 90% of dissolved methane could be recoverable at a retention time of 11 min and a mixing speed of 800 rpm. Including this recovered methane, the total methane recovered from anaerobic treatment of wastewater reached 54% of influent COD (Gu *et al.*, 2017).

The suspended solids concentration in AMBBR effluent was found to be as low as 38 mg/L VSS, corresponding to an observed sludge yield of 0.1 g VSS/g COD removed. The carbon distribution in the AMBBR process at steady state further indicates that almost 15% of the influent carbon was assimilated into biomass, 19% remained in the AMBBR effluent, and 28% was converted to methane with 10% in dissolved form (Figure 3.13).

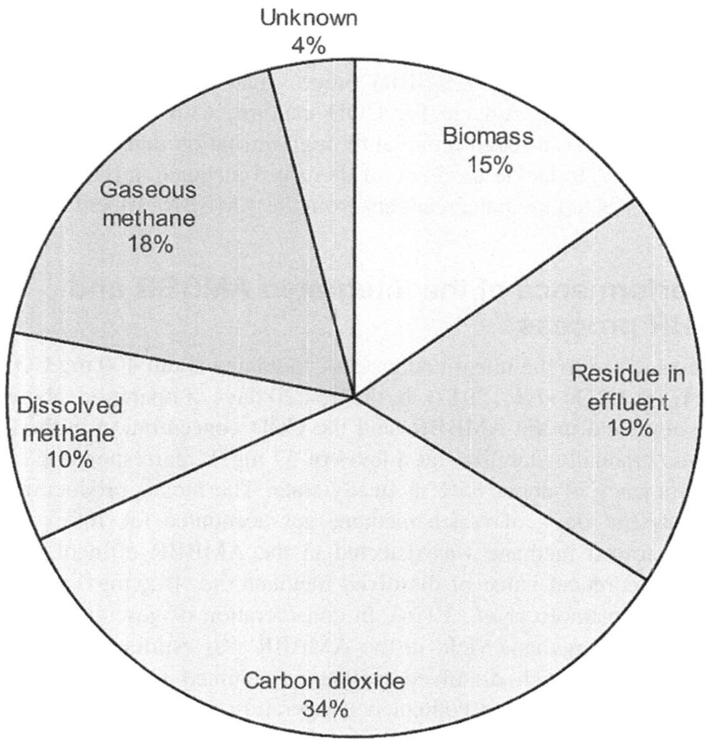

Figure 3.13 Carbon distributions in the AMBBR process. Data from Gu *et al.* (2017).

In the integrated process (Figure 3.12), an IFAS-SBR unit was used to serve as the B-stage for nitrogen removal. It was observed that 87% of TN was removed in the period from day 10 to day 55, during which the COD/N ratio in the influent to the IFAS-SBR was greater than 1.9. However, with an improved A-stage performance, the B-stage influent COD/N ratio gradually decreased from day 55 onward, which led to a gradual rise of nitrate concentration in the effluent (e.g. about 20 mg N/L). This in turn indicated a deterioration of nitrite shunt and denitritation.

During the initial period of operation, most of the influent ammonium was oxidized to nitrite in the oxic phases with an NAR of 80%. The produced nitrite was subsequently removed in the anoxic phases. This indicates that nitritation-denitritation was established in the initial period of operation. On day 48, nitrate concentration increased to about 3 mg N/L by the end of each oxic phase, while nitrite could not be fully denitrified in the anoxic phases due to the decrease in the COD/N ratio in the feed to the B-stage. Microbial analysis revealed that the abundance of denitrifiers in both the suspended biomass and biofilms declined from day 20 to 47, implying the occurrence of incomplete denitrification. On day 105, almost all ammonium was converted to nitrate in the oxic phases with limited nitrite accumulation; however, nitrite and nitrate could not be totally denitrified in the anoxic phases, likely due to the low COD concentration in the step-feed reactor (e.g. COD/N ratio of around 1.5). It was concurrently observed that the gene copy number of denitrifiers in the suspended biomass tended to further decline to 5.8×10^{10} copies/g VSS. By contrast, the abundances of AOB and *Nitrospira* increased to 3.9×10^{10} and 1.2×10^{11} copies/g VSS, respectively. In the fixed biofilms, *Nitrospira* abundance stood at the level of 1.4×10^{10} copies/g SS, while the denitrifiers experienced a clear decline in abundance. This all suggests that the COD/N ratio in the influent to the B-stage IFAS-SBR, to a great extent, determined nitrogen removal and the microbial community structure.

Alternating anoxic/oxic conditions have been proven effective for NOB repression against AOB (Ge *et al.*, 2014; Kornaros *et al.*, 2008; Wang *et al.*, 2019; Yang & Yang, 2011). In the A-B process described in Figure 3.12, successful nitrite accumulation was achievable in the oxic phases under alternating anoxic/oxic conditions when sufficient COD was available for denitrification. However, nitrite shunt was gradually deteriorated when influent COD concentration to the B-stage decreased with improved performance of the A-stage AMBBR. In fact, the remaining nitrite in the anoxic phases eventually promoted the growth of NOB in the subsequent oxic phases (Gu *et al.*, 2018b). This was also supported by the fact that the gene copy number of NOB on day 109 in the presence of inadequate COD for denitrification was almost 2.3 times higher than that on day 20 with enough COD for denitrification. Thus, it appears that the alternating anoxic/oxic conditions appear to be necessary but are not sufficient to maintain a sustainable stable nitrite shunt.

3.6.3 Cost-benefit analysis

3.6.3.1 Sludge production and recovery of dissolved methane

It is known that about 0.3–0.5 g dry biomass is generated in the CAS process when removing 1 g of COD (Wan *et al.*, 2016). Further, the observed growth yield of anaerobic sludge has been found to be in the range of 0.03–0.2 g SS/g COD (Lettinga *et al.*, 1983). In the integrated A-B process with AMBBR as the lead, the observed sludge yield was determined to be about 0.1 g dry biomass/g COD, indicating 75% less sludge production when compared with the CAS process. Obviously, adopting an anaerobic unit as the lead in the A-B process has the advantages of improved energy recovery directly from municipal wastewater, with reduced energy consumption and minimized waste sludge generation (Gu *et al.*, 2017).

Today, the wastewater–energy–greenhouse gas nexus in WWTPs receives increasing attention for its impact on in-plant energy recovery and global climate change. In the integrated process with an anaerobic unit as the lead, dissolved methane becomes a hurdle for large-scale application. As shown above, dissolved methane may account for about 35% of the overall methane production in an AMBBR. It should be noted that the solubility of methane is related to temperature, e.g. the issue of dissolved methane would be more serious in cold regions. For example, 50% of methane produced in an AnMBR was in dissolved form at 15°C (Smith *et al.*, 2013). Therefore, proper recovery or removal of dissolved methane is required for large-scale applications of anaerobic treatment processes to secure economic viability and environmental sustainability in WWTPs. As discussed above, the flash chamber developed by Gu *et al.* (2017) was able to achieve 90% dissolved methane recovery with 1.7 mg CH_4/L in AMBBR effluent. Currently, several strategies have been investigated for the removal and recovery of dissolved methane from anaerobic effluents, e.g. biological oxidation, air stripping, degassing membrane etc. (Bandara *et al.*, 2011; Cookney *et al.*, 2016; Matsuura *et al.*, 2010; Matsuura *et al.*, 2015). Matsuura *et al.* (2015) developed a two-stage down-flow hanging sponge (DHS) reactor in which the majority of dissolved methane was recovered in a first DHS reactor by air gasification and the residual dissolved methane was further removed in a second DHS reactor via biological oxidation. However, the DHS reactor had a large footprint due to the long HRT required. Although a degassing membrane showed a high methane recovery rate, its application was seriously challenged by pore wetting and clogging (Crone *et al.*, 2016). Apparently, the flash chamber may offer another option for dissolved methane removal or recovery with the advantages of its simple structure, easy upscaling, simple operation and maintenance, small footprint, high efficiency and relatively low operation cost.

3.6.3.2 Energy gain in integrated anaerobic and nitritation-denitritation process

AD has been commonly used to recover energy from WAS produced in the CAS process, e.g. about 33% of influent COD could be recovered as methane gas via AD (Wan et al., 2016). In the integrated process with AMBBR as the leading A-stage, the recovery rate was increased to 55%. The potential electrical energy from the generated methane was calculated as 0.28 kWh/m^3 wastewater treated, using Equation 3.1. On the other hand, the energy consumption by the flash chamber was estimated at 0.12 kWh/m^3 (Gu *et al.*, 2017).

Apart from the energy recovery at A-stage, substantial reduction in sludge production could also be achievable, together with remarkable reduction in aeration-associated energy consumption at the B-stage. In present processes, the energy consumed by sludge treatment and aeration account for 30% and 50% of the overall power usage, respectively (Henderson, 2002). For a typical energy consumption of 0.47 kWh/m^3 in the CAS process (Goldstein & Smith, 2002; Zhou *et al.*, 2013), the energy savings due to reduced aeration and sludge production could reach 0.24 kWh/m^3 and 0.11 kWh/m^3, respectively, i.e. a total of 0.35 kWh/m^3. This suggests that the energy gain achievable in the integrated process (Figure 3.12) would be around 0.51 kWh/m^3, which is more than enough to offset the energy usage in the CAS process. Thus, it appears possible and viable to obtain energy-neutral or even positive operation in WWTPs by adopting the integrated A-B process illustrated in Figure 3.12.

3.6.4 Energy recovery versus nitrogen removal via nitritation-denitritation

To enhance the energy efficiency of the A-B process, direct COD capture and conversion to methane gas should theoretically be maximized at the A-stage. However, this may create an unfavourable situation at the B-stage due to the shortage of COD for denitritation. Theoretically, 1.71 g of COD is needed for complete conversion of 1 g of NO_2^--N to nitrogen gas. Given the COD capture efficiency of 68% at the A-stage, the COD entering the B-stage is inadequate to sustain denitritation. Therefore, a compromise solution for both energy recovery and efficient denitritation has been proposed to address this emerging challenge (Gu *et al.*, 2018b).

3.6.5 New A-B process configuration with bypass of influent

As discussed above, the more COD harvested in the A-stage AMBBR, the less COD was available for denitritation in the B-stage IFAS-SBR. As such, the performance of denitritation was eventually compromised. It is sensible to consider that the COD

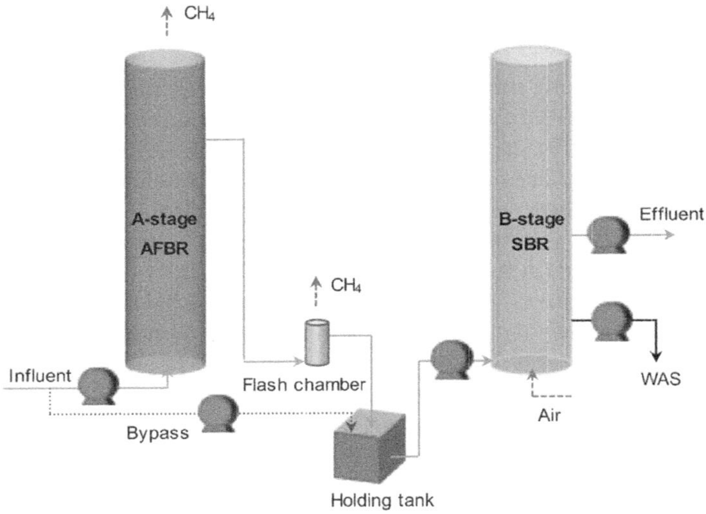

Figure 3.14 Schematic of A-B process integrating anaerobic treatment and nitritation-denitritation with bypass.

required for nitrogen removal at the B-stage should determine to what degree the influent COD could be captured through the A-stage. Therefore, the operation of the integrated A-B process (Figure 3.12) should be optimized in terms of both energy recovery and nitrogen removal in a holistic manner. To face this situation, an anerobic fixed bed reactor (AFBR) was adopted as the A-stage with an SBR being the B-stage, with a part of influent municipal wastewater directed to mix with AFBR effluent prior to entering the B-stage (Figure 3.14).

3.6.6 General performance of the anaerobic treatment and nitritation-denitritation process with bypass

The integrated A-B process was operated in two phases, i.e. phase 1 without bypassed flow and phase 2 with bypassed flow. In the AFBR (A-stage), about 80% of influent COD was removed and 74 mg/L COD remained in the anaerobic effluent, with a sludge yield of 0.09 g biomass/g COD removed (Gu *et al.*, 2018b). Meanwhile, 73.87 L of biogas was produced from one cubic meter of wastewater treated, of which 67% was methane. The COD concentration in the B-stage effluent stabilized at around 10 mg/L during the entire operation period, suggesting that no negative influence was observed on COD removal at the B-stage with the implementation of bypassing 19% of the influent to the B-stage.

During phase 1, the majority of ammonium was oxidized to nitrite with an obvious nitrite shunt, corresponding to an NAR of above 86%. From day 12 to day 30, TN removal reached about 78% with an influent COD/N ratio higher

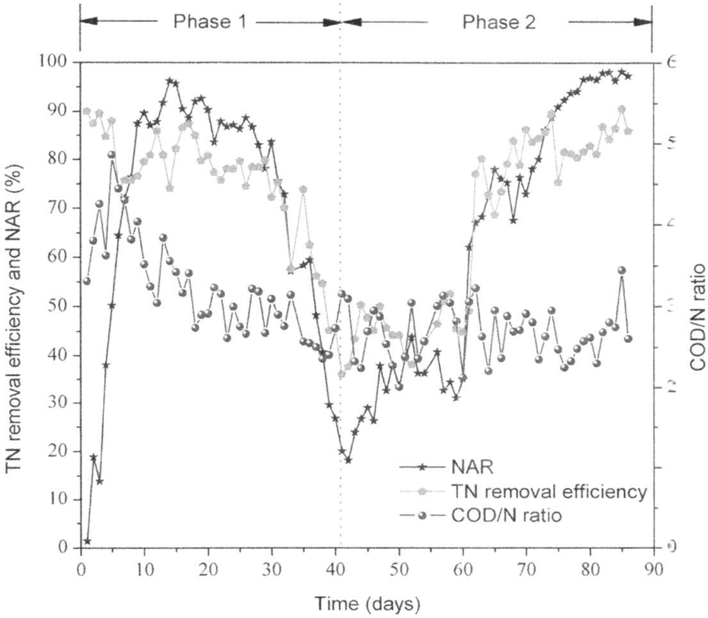

Figure 3.15 TN removal efficiency, NAR and COD/N ratio in a B-stage SBR. Adapted from Gu *et al.* (2018b), with permission.

than 2.9 (Figure 3.15). However, with the decrease in the COD/N ratio of the A-stage effluent, nitrate started to accumulate in the B-stage effluent, indicating the possible deterioration of the nitrite shunt and incomplete denitritation because of the shortage of COD supply by the end of phase 1. To address this issue, the integrated A-B process was modified by bypassing 19% of influent wastewater, which was further blended with the AFBR effluent (Figure 3.14). The blended stream with a COD/N ratio of 2.5 to 3 was then fed to the B-stage for nitrogen removal through nitritation-denitritation. After a 20-day operation, the nitrate concentration in the B-stage effluent gradually reduced from about 15 mg N/L to 1.5 mg N/L. In the meantime, NAR and TN removal were restored to levels of 97% and 85%, respectively.

3.6.7 Effective NOB repression in the nitritation-denitritation process

NOB repression could be reflected by specific AOB activity ($q_{NH_4^+}$) versus that for NOB ($q_{NO_2^-}$). In phase 1, $q_{NH_4^+}$ was only about 1.3 times that of $q_{NO_2^-}$. In phase 2, $q_{NO_2^-}$ decreased significantly from 0.55 g N/(g VSS d) to 0.17 g N/(g VSS d), while $q_{NH_4^+}$ remained relatively constant after the bypass was introduced

Table 3.3 Specific AOB and NOB activities in Phase 1 and Phase 2.

	$q_{NH_4^+}$, g N/ (g VSS · d)	$q_{NO_2^-}$, g N/ (g VSS · d)
Phase 1	0.74	0.55
Phase 2	0.67	0.17

(Table 3.3). Obviously, the bypass flow helped to suppress NOB activity but without compromising AOB activity. As such, a stable nitrite shunt was established.

The profiles of the nitrogen cycle-related microbial species in the B-stage showed that the abundance of denitrifiers decreased in phase 1 and subsequently increased 4-fold in phase 2 after implementation of the bypassing flow. This trend was consistent with the observed TN removal (Figure 3.15). Meanwhile, it was found that AOB abundance was comparable in the two phases, while NOB abundance declined significantly after implementation of the bypassed flow in phase 2. During one cycle of operation, ammonium was mainly converted to nitrite in each oxic sub-cycle with the nitrate produced being lower than 0.5 mg N/L, while the nitrite was removed in subsequent anoxic sub-cycles with the presence of the bypassed COD in the B-stage influent. This indicated the establishment of excellent denitritation-denitritation in Phase 2 with the bypassed influent flow.

Currently, different operation strategies have been reported for NOB repression in mainstream municipal wastewater treatment, e.g. short SRT, DO concentration, alternating anoxic/oxic conditions etc. However, as shown above, control of any individual parameter alone could not lead to effective NOB repression in A-B processes where the majority of COD was harvested at the A-stage. To achieve sustainable nitritation-denitritation, nitrite should be timely removed via denitritation without accumulation since any accumulated nitrite may trigger the growth of NOB, resulting in nitrate accumulation in the oxic environment (Gu *et al.*, 2018b; Xu *et al.*, 2017). Therefore, maintaining sustainable and stable nitritation-denitritation with the A-B process appears challenging (Figure 3.12). To minimize nitrite conversion to nitrate at the B-stage, a sufficient amount of COD should be available to denitrify the nitrite in a timely manner.

3.7 INTEGRATED ANAEROBIC AND TWO-STAGE DEAMMONIFICATION PROCESS

3.7.1 Process configuration

Anammox has been regarded as an energy-efficient alternative for nitrogen removal (Strous *et al.*, 1998). Currently, anammox technology has been successfully applied for treating high-strength wastewater, e.g. sludge digestion liquor (van der Star *et al.*, 2007; Wang *et al.*, 2017a; Xu *et al.*, 2018), landfill leachate (Anfruns *et al.*, 2013; Sun *et al.*, 2016), etc. However, it is still challenging to apply anammox in low-strength municipal wastewater treatment due to the difficulties in sustaining stable nitrite shunt, competition between anammox bacteria and denitrifiers, and effective retention of anammox bacteria, etc. (Hendrickx *et al.*, 2012; Xu *et al.*, 2015). Two-stage processes have been investigated for partial nitritation and anammox, allowing optimization of partial nitritation and anammox in a separate manner (Ma *et al.*, 2011; Regmi *et al.*, 2015; Regmi *et al.*, 2016). However, it is still difficult to generate a stream with a suitable nitrite to ammonium ratio for mainstream anammox. Real time control based on on-line sensing of ammonium, nitrite, nitrate, pH, etc, has been proposed, but the whole system seemed highly complicated (Jiang *et al.*, 2018; Langone *et al.*, 2016; Regmi *et al.*, 2015). To address these issues, an integrated process, called the A-2B process, was developed, as illustrated in Figure 3.16.

As shown in Figure 3.16, an AFBR served as the A-stage for direct COD capture and biogas production, followed by a flash chamber for dissolved methane recovery, an SBR as the B1-stage and a moving bed biofilm reactor (MBBR) as the B2-stage for nitrite production and the anammox process, respectively. In this process, the influent was partially bypassed to the B1-stage SBR through Bypass 1 after blending with A-stage effluent in a holding tank. The SBR was operated in 4 sub-phases consisting of step-feeding, anoxic (20 min) and oxic (13 min) phases with DO concentration of 1.5 mg/L in oxic phases and with an SRT maintained at 4 days. Thereafter, a portion of the A-stage effluent mainly containing ammonium was diverted via Bypass 2 to blend with the SBR effluent, and further fed to the MBBR of the B2-stage for nitrogen removal via anammox. Cubic sponges were used as biocarriers in the MBBR with a packing ratio of 50%. The ratio of Bypass 1 was fixed at 16% of the influent flow, while that of Bypass 2 was gradually adjusted to 22% (phase 2 from day 74 onwards).

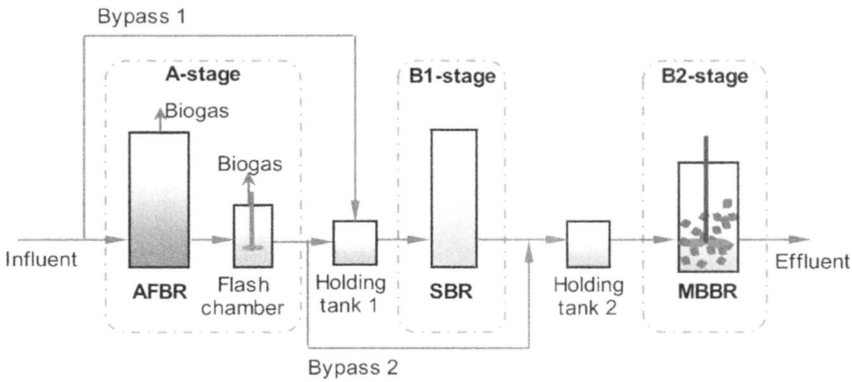

Figure 3.16 Schematic of the A-2B process. Adapted from Gu *et al.* (2018a), with permission.

3.7.2 Performance of the integrated anaerobic and two-stage deammonification process

The integrated process was operated with synthetic municipal wastewater for 125 days. The COD concentration was reduced from 384 mg/L to nearly 29 mg/L in the A-stage AFBR, with a methane gas yield of 0.27 L/g COD removed. About 41% of total methane produced was in the dissolved form, about 90% of which was recovered in the flash chamber. The sludge yield in the A-stage was calculated as 0.06 g dry biomass/g COD removed. Due to the nature of anaerobic reactions, no ammonium removal was observed.

In the B1-stage SBR, the influent ammonium was mostly oxidized to nitrite with a clear nitrite shunt. Due to reduced COD concentration in the AFBR effluent, the NO_2^--N concentration in the B-stage effluent gradually stabilized at 15 mg/L, while the nitrate concentration remained at a level below 5 mg N/L. The specific oxygen uptake rate (SOUR) of AOB was determined to be 1.67 g O_2/ (g VSS d), which was 3.6 times higher than that of NOB. Obviously, a stable nitrite shunt was established in the B-stage SBR. In fact, the alternating anoxic/oxic conditions together with the aggressive SRT control have been proven effective in NOB repression (Gu *et al.*, 2018b; Xu *et al.*, 2017).

The concentration profiles of NH_4^+-N, NO_2^--N, NO_3^--N and COD in the anammox MBBR are shown in Figure 3.17. The B2-stage MBBR was fed with a

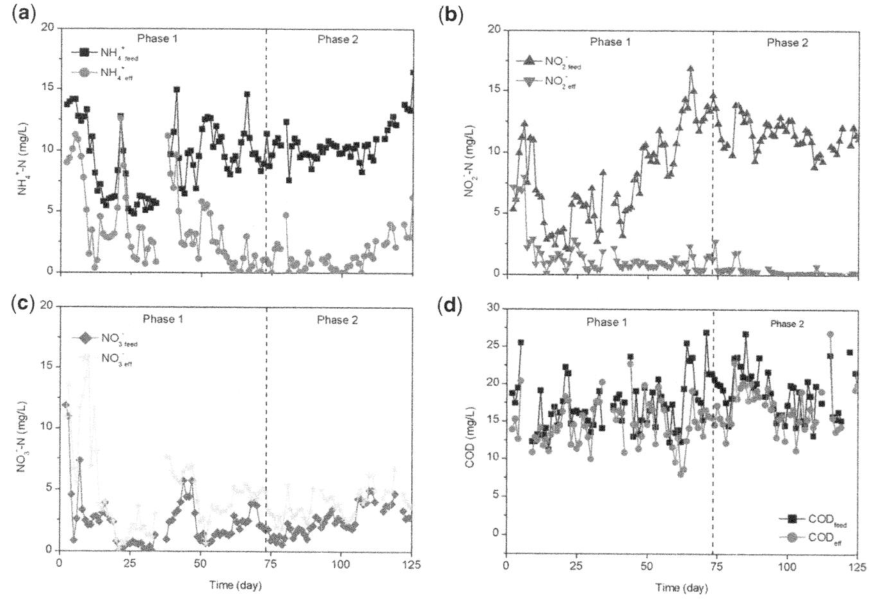

Figure 3.17 Profiles of nitrogen species and COD in the B2-stage MBBR. Adapted from Gu *et al.* (2018a), with permission.

mixture of the B1-stage SBR effluent containing nitrite and the A-stage AFBR effluent with ammonium. As can be seen in Figure 3.17, with the gradual increase of nitrite concentration after 25 days of operation, simultaneous reduction of NH_4^+-N and NO_2^--N was observed in phase 1. Subsequently, the ratio of Bypass 2 was fixed at 22% in phase 2, which created an ideal ratio between the concentrations of NO_2^--N and NH_4^+-N at around 1.1 for anammox. Accordingly, the NH_4^+-N and NO_2^--N concentrations were reduced to below 2 mg N/L and 1 mg N/L on average, respectively, in the anammox MBBR effluent. Meanwhile, very little nitrate was generated with nearly no COD consumption, suggesting that the NH_4^+-N and NO_2^--N losses in the MBBR were due to the anammox activity. At the steady state with a nitrogen loading rate of 0.12 kg N/(m^3 d), 74% of nitrogen could be removed in the anammox MBBR.

The NO_2^-/NH_4^+ and NO_3^-/NH_4^+ ratios determined by *in situ* batch tests were found to be similar to those reported in anammox processes as shown in Table 3.4. It was further found that the ratios of $NO^-_{2\,removed}/NH^+_{4\,removed}$ ratio and $NO^-_{3\,produced}/NH^+_{4\,removed}$ gradually approached the theoretical values after 45 days of operation, suggesting a fast start-up of the anammox MBBR.

Compared to one-stage deammonification, two-stage deammonification segregates partial nitritation and anammox into two consecutive steps in separate reactors. As such, the inhibition on anammox bacteria because of exposure in oxygen can be avoided (Li *et al.*, 2011), and AOB activity can be optimized by applying higher DO concentration in a partial nitritation reactor (Regmi *et al.*, 2014; Regmi *et al.*, 2015). More importantly, the SRTs of the nitrifying sludge and anammox can be separated, allowing concurrent NOB repression and anammox retention (Laureni *et al.*, 2019). As an equimolar ratio of NH_4^+-N and NO_2^--N must be secured for anammox, the operation strategy adopted was biological, converting almost half of the ammonium to nitrite (Ma *et al.*, 2011; Wyffels *et al.*, 2004). However, it is rather challenging to maintain this ratio during long-term operation. By contrast, the A-2B process offers a great opportunity for easy system control.

3.7.3 Microbial community in the B2-stage

16S rRNA gene sequencing analysis indicated that *Candidatus* Kuenenia was dominant in the biofilms with a relative abundance of 14.19%, while the other

Table 3.4 Stoichiometric coefficients of anammox reported in the literature.

Reactor	NO_2^-/NH_4^+ Ratio	NO_3^-/NH_4^+ Ratio	References
anammox MBBR	1.27 ± 0.05	0.28 ± 0.03	Gu *et al.* (2018a)
MBR	1.2	0.21	Lotti *et al.* (2014)
SBR	1.32	0.26	Strous *et al.* (1998)

two genera affiliated with anammox were also identified as *Candidatus* Brocadia and *Candidatus* Anammoxoglobus. For the nitrifying bacteria, *Nitrolancea,* which can tolerate nitrite at high concentration, was the dominant NOB with an abundance of 1.52%, while *Nitrospira* and *Nitrobacter,* which are commonly found in WWTPs (Whang *et al.*, 2009), had abundances lower than 0.94%. On the other hand, *Nitrosomonas* only accounted for 0.14%, and other ammonia oxidizing bacteria (e.g. *Nitrosococcus* and *Nitrosospira*) were nearly undetectable.

3.7.4 Mass balance and cost-benefit analysis

A mass balance of the A-2B process revealed that 58% of influent COD was recovered as methane through the A-stage AFBR, with 11% of influent COD assimilated into wasted sludge (Figure 3.18). The observed sludge yield was estimated to be 0.11 g VSS/g COD removed, whereas the CAS process normally converts 30–50% of influent COD to sludge, with a typical sludge yield of 0.4 g VSS/g COD removed. Currently, a large amount of WAS is generated in WWTPs with the CAS as a core unit process. For example, dry sludge production reached 8 million tons in the USA and almost 11 million tons in China in 2010. Although AD of WAS has been adopted to reduce the volume of WAS and recover energy, overall energy efficiency appeared to be very low. For example, merely 30–50% of VSS could be successfully processed, and the digestate still requires further disposal.

The overall energy efficiency of current CAS-based WWTPs has been reported to be as low as 20–50% (Zhou *et al.*, 2013). Indeed, current wastewater treatment philosophy is established on the basis of biological oxidation, by which influent COD is oxidized to carbon dioxide at a high energy cost with huge amounts of WAS produced. It has been shown that more energy could be recoverable from wastewater through direct COD capture instead of bio-oxidation. For example, anaerobic treatment of municipal wastewater offers an alternative towards

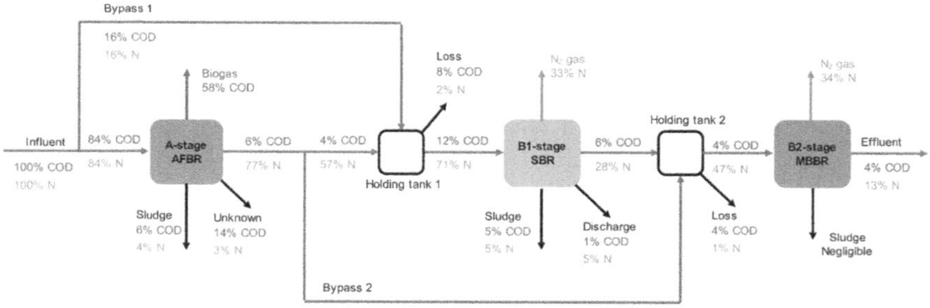

Figure 3.18 COD and nitrogen flows in the A-2B process. Adapted from Gu *et al.* (2018a), with permission.

improved energy efficiency (McCarty *et al.*, 2011). In this A-2B process, a significant amount of COD in the wastewater was captured to produce biogas directly with minimized production of waste sludge. On the other hand, TN removal reached 87% in the A-2B process, in which 33% and 34% were removed via nitritation-denitritation in the B1-stage SBR and the anammox B1-stage MBBR, respectively. As such, energy consumption for nitrogen removal could be cut by around 47% as compared with full nitrification and denitrification. Therefore, the A-B process in this configuration is practically feasible and economically viable for energy-neutral or even energy-positive WWTPs operation.

3.8 INTEGRATED ANAEROBIC AND SINGLE-STAGE DEAMMONIFICATION PROCESS

3.8.1 Process configuration

Mainstream deammonification can be realized either as a single-stage or two-stage process (Gu *et al.*, 2018a; Regmi *et al.*, 2015). Single-stage deammonification has the advantages of low capital cost on infrastructure and a simple process configuration, and has thereby gained increasing interest. However, many challenges still remain unresolved for its large-scale application in municipal wastewater treatment, e.g. NOB repression, microbial competition, retention of anammox bacteria, etc. The low DO strategy has been widely adopted to repress NOB activity (Laureni *et al.*, 2016; Zheng *et al.*, 2019); however, unsuccessful NOB repression has been widely reported, which might be because of the variations in oxygen affinities among different NOB genera (Regmi *et al.*, 2014; Strous *et al.*, 1999). Moreover, it has been found that DO at low concentrations may inhibit AOB activity as well and eventually result in nitrite limitation for anammox bacteria (Lotti *et al.*, 2015).

As discussed above, a step-feed SBR exhibited a great potential for sustaining stable partial nitritation (Gu *et al.*, 2018a). However, a SBR is a discontinuous process which is operated in a temporal mode. Indeed, the temporal operation of an SBR could be realized within the spatial operation of a step-feed reactor. Therefore, combining partial nitritation and anammox seems highly feasible in a single-stage step-feed reactor, which might be able to achieve a more stable deammonification. As such, an integrated process was developed as shown in Figure 3.19. Briefly, an AFBR was adopted as the leading A-stage for direct biogas production from influent COD, and the effluent was then fed into a step-feed reactor developed for deammonification. The step-feed reactor was divided into 5 anoxic and 4 oxic chambers, while the influent was equally pumped into the first 4 anoxic chambers, i.e. A1–A4, while all the anoxic chambers were packed with anammox biocarriers at a filling ratio of 50% (Figure 3.19). The DO concentration was controlled in the range of 1.2–1.5 mg/L

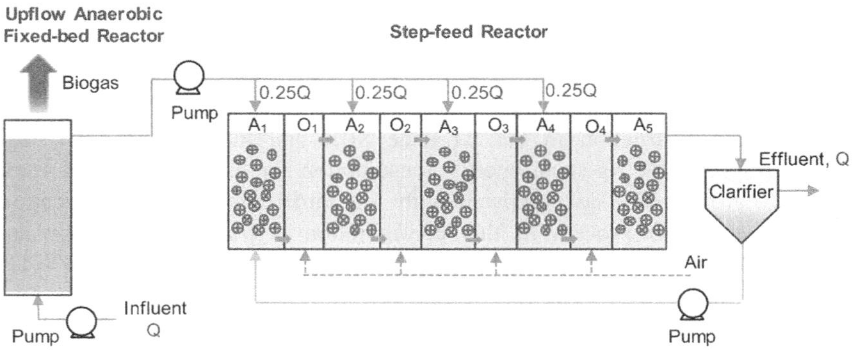

Figure 3.19 Schematic of the A-B process integrating single-stage deammonification. Adapted from Gu *et al.* (2019), with permission.

in the oxic chambers, and the HRTs of the AFBR and step-feed reactor were controlled at about 6 and 5.8 hours, respectively.

3.8.2 Process performance

The integrated process was fed with a synthetic wastewater containing 400 mg/L of COD and 45 mg N/L of ammonium. The COD concentration was reduced to about 19 mg/L in the AFBR, and the observed sludge yield was determined to be about 0.05 g VSS/g COD, which was significantly low when compared with the typical value of 0.45 g VSS/g COD in the CAS process. This indicated that sludge production in the integrated process could be reduced by about 89% as compared with the CAS process. Meanwhile, 48.8% of the removed COD was directly channeled to gaseous methane.

As the majority of the COD in the influent was harvested in the A-stage AFBR, a COD/N ratio of 0.44 was generated for subsequent anammox in the B-stage. The ammonium and nitrite concentrations in the B-stage effluent gradually decreased and finally stabilized at around 3.2 and 3.0 mg N/L, respectively (Figure 3.20). Meanwhile, the nitrate concentration was found to be below 5 mg N/L. It should also be noted that after 10 days of operation, about 60% of TN was already removed, suggesting a fast start-up of the single-stage deammonification process.

3.8.3 Nitrogen conversion in the step-feed mainstream deammonification process

The nitrogen balance in the steady state step-feed reactor indicated that 0.7 g NH_4^+-N/d was removed in the reactor, equivalent to 96% removal of NH_4^+-N, of which 51.6% occurred in the oxic chambers, whilst the rest was in the anoxic

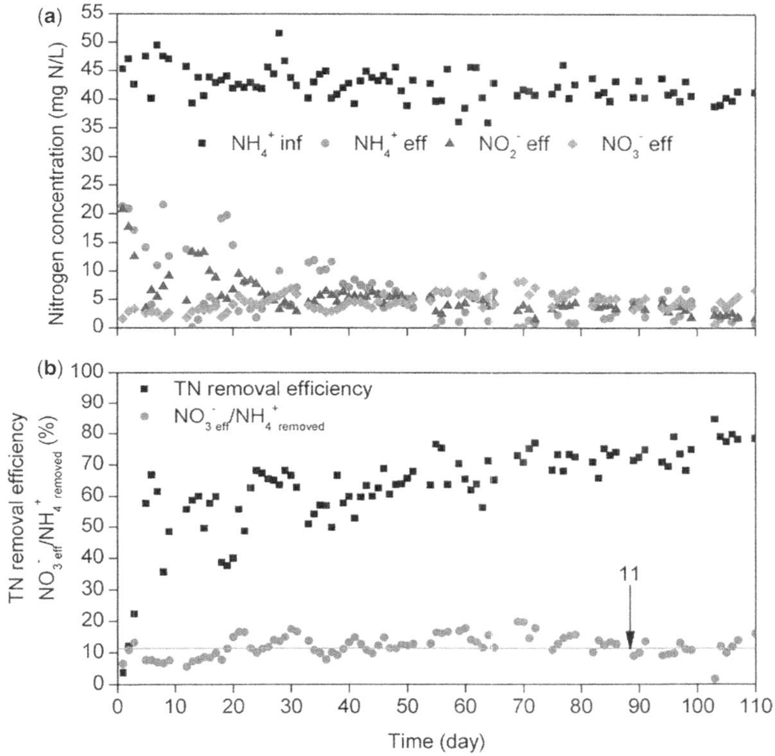

Figure 3.20 Profiles of nitrogen species (a) and TN removal efficiency and NO_3^--N_{eff}/NH_4^+-$N_{removed}$ ratio (b) in a B-stage step-feed reactor. Adapted from Gu *et al.* (2019), with permission.

chambers. In the oxic chambers (i.e. O_1–O_4), ammonium removal was mainly due to its oxidation to nitrite with a conversion ratio of 64.3–84.8%, and nitrate production only made up 5.16% of the total ammonium removed. This suggests the successful repression of NOB in the step-feed reactor.

In the oxic chambers, about half of the ammonium was converted to nitrite, which was carried over to the next anoxic chamber and removed via the anammox pathway together with the presence of ammonium in the influent of the step-feed reactor. The TN removal rate of 0.59 g/d was achieved in the step-feed reactor, of which 87.1% was removed in the anoxic chambers, with the concurrent removal of ammonium and nitrite. With sustainable nitrite production in the oxic chambers, 16.4–24.8% of TN was removed in the subsequent anoxic chambers, i.e. A_2–A_5. However, the TN removal in the A_1 chamber was low, likely due to the fact of the nitrite limitation for anammox bacteria.

3.8.4 NOB repression in a step-feed reactor

Repression of NOB is crucial to achieve a stable nitrite shunt, which is a prerequisite to sustain anammox. However, sustainable partial nitritation in the mainstream still remains as a major hurdle for deammonification. The NO_3^--N_{eff}/NH_4^+-$N_{removed}$ ratio, an indicator reflecting the degree of NOB repression (Lotti *et al.*, 2015), obtained in this step-feed reactor was comparable to its theoretical value of 0.11 (Figure 3.20). This in turn suggests that effective NOB repression was realized in the step-feed reactor. In fact, AOB activity was found to be nearly 6.3 times higher than NOB activity. Microbial analysis further indicated that the relative abundance of AOB rose by a factor of 8 along with operation. In contrast, the relative abundance of *Nitrospira* was relatively stable, while *Nitrobacter* showed a clear decreasing trend in abundance.

Different strategies have been proposed to repress NOB activity in mainstream deammonification processes, e.g. free nitrous acid (FNA) (Wang *et al.*, 2014), free ammonia (FA) (Wang, 2017; Wang *et al.*, 2017b) etc. Akaboci *et al.* (2018) reported NOB repression by coupling the activities of AOB and anammox bacteria, where adjustment of the nitrogen loading rate (NLR) was needed to match the actual anammox activity. These strategies seem complicated in execution. In the step-feed reactor, suspended sludge continuously experienced the alternative anoxic/oxic conditions in a spatial fashion. The effective NOB repression could be caused by the anoxic disturbances on the nitrifying sludge (Bournazou *et al.*, 2013; Kornaros *et al.*, 2008; Kornaros *et al.*, 2010). Moreover, the nitrite limitation at the beginning of the oxic phases also provided a favourable condition for NOB repression due to the fact that the accumulation of nitrite in extended oxic phases promotes the growth and activity of NOB (Gu *et al.*, 2018b; Xu *et al.*, 2017).

3.8.5 Anammox in the step-feed mainstream deammonification process

Maintaining high anammox activity is also critical for the single-stage deammonification process in long-term operation. The specific anammox activity in biofilms was determined to be 0.42 g N/(g VSS d), which was much more robust than that reported in the literature (Lotti *et al.*, 2015; Wang *et al.*, 2018). It should be noted that anammox bacteria would be outcompeted by heterotrophic denitrifiers with the presence of sufficient COD (Leal *et al.*, 2016). Therefore, the decrease in activity or even washout of anammox bacteria has been reported in treatment of wastewater with high COD/N ratio (Kuenen, 2008). In the A-B process, the AFBR was purposely designed as the A-stage for enhanced methane production with minimized sludge generation and effluent with a reduced COD/N ratio to the B-stage mainstream deammonification. The results showed that a low COD/N ratio of 0.44 could be obtained, which was favourable for anammox bacteria (Xu *et al.*, 2015). The robust anammox activity in turn could be partially explained by the unique process configuration in the step-feed reactor.

3.8.6 Cost-benefit analysis

Currently, the CAS process, being widely adopted for municipal wastewater treatment in WWTPs, has been subject to increasing concerns about its low energy efficiency and WAS production. In the Ulu Pandan Water Reclamation Plant (WRP), Singapore, and Gaobeidian WWTP, China – both with the CAS as a core treatment unit – only about 34% and 31% of energy efficiency were realized, respectively, even equipped with AD for energy generation from WAS (Cao, 2011; Zhou *et al.*, 2013). Apart from low energy efficiency, the treatment and disposal of WAS has become a global issue because landfill is no longer an acceptable option for many reasons, while incineration has been regarded as a costly and non-environmentally friendly approach with the generation of ashes and harmful gases (Liu *et al.*, 2018). Due to the rapid growth of wastewater treatment capacity, the quantity of WAS generated will reach 90 million tons by 2030 in China. Some theoretical analysis has shown that further optimizing the CAS process could not make significant improvements in terms of energy efficiency and sludge reduction (Wan *et al.*, 2016). In fact, it is highly possible to achieve energy-efficient wastewater treatment by adopting the integrated process AFBR and step-fed process as described in Figure 3.19.

The total in-plant energy consumption of the CAS process was 0.44 kWh/m^3, while 0.15 kWh/m^3 of energy was recoverable through AD of WAS, i.e. a net energy consumption of 0.29 kWh/m^3. However, if the integrated A-B process (Figure 3.19) was adopted, the energy consumption for sludge treatment would be 84% less, compared with that in the CAS process. In addition, the energy consumption associated with fine bubble aeration in the CAS process was 0.23 kWh/m^3, of which 0.14 kWh/m^3 and 0.09 kWh/m^3 were consumed for COD oxidation and full nitrification, respectively (Cao, 2011). By contrast, the energy for COD oxidation could be totally saved in the integrated process as most of influent COD was channeled to methane in the AFBR. Furthermore, in the deammonification process (i.e. the B-stage), only half of the influent ammonium was oxidized to nitrite. Theoretically, the oxygen demand for nitritation was 3.43 g O_2/g ammonium-N. As such, the aeration-associated energy consumption could be calculated as 0.03 kWh/m^3, if the oxygen transfer efficiency was assumed to be 3 kg O_2/kWh (Crawford & Sandino, 2010).

In this integrated A-B process, 48.8% of the COD was directly captured in the form of methane gas in the AFBR, which was much higher than the 17.9% in CAS process. The gaseous methane generated in the integrated A-B process was able to produce 0.37 kWh/m^3 recoverable electrical energy against 0.15 kWh/m^3 in the CAS process. As a result, a net energy gain was estimated at 0.19 kWh/m^3 under the assumptions that the other energy consumption (e.g. pumping) was the same as those in the CAS process. Consequently, the A-B process presented in Figure 3.19 may offer a viable engineering solution towards sustainable mainstream deammonification with reduced energy consumption, enhanced energy recovery and minimized sludge generation.

3.9 CONCLUSIONS

The energy consumption in WWTPs adopting the CAS process has been reported to be 0.47 kWh/m^3 on average, of which aeration and sludge treatment can account for more than 50% and 20% of the energy consumption, respectively (Goldstein & Smith, 2002). It has been argued that any significant reduction in in-plant energy consumption is unachievable by further process optimization in the current CAS process. In fact, municipal wastewater is a source rich in various resources including water, energy and various other resources. Facing such a situation, the concept of the A-B process has been naturally developed, based on the philosophy of COD capture for energy production at the A-stage, followed by energy efficient nitrogen removal at the B-stage. Improved energy efficiency can be achieved in the various A-B processes that have been clearly presented in this chapter.

In A-B processes with an anaerobic unit as the lead, COD was directly converted to methane gas without production of WAS, while in processes with CEPT or HRAS as the lead, energy can be recovered from produced sludge via AD. This implies that the A-B processes with CEPT or HRAS as an A-stage need to be more complicated in process configuration and control. It appears that the A-B processes discussed in this chapter may open a new window on energy neutral or even energy positive municipal wastewater reclamation.

REFERENCES

Akaboci T. R.V., Gich F., Ruscalleda M., Balaguer M. D. and Colprim J. (2018). Assessment of operational conditions towards mainstream partial nitritation-anammox stability at moderate to low temperature: reactor performance and bacterial community. *Chemical Engineering Journal*, **350**, 192–200.

Anfruns A., Gabarró J., Gonzalez-Olmos R., Puig S., Balaguer M. and Colprim J. (2013). Coupling anammox and advanced oxidation-based technologies for mature landfill leachate treatment. *Journal of Hazardous Materials*, **258**, 27–34.

Bandara W. M., Satoh H., Sasakawa M., Nakahara Y., Takahashi M. and Okabe S. (2011). Removal of residual dissolved methane gas in an upflow anaerobic sludge blanket reactor treating low-strength wastewater at low temperature with degassing membrane. *Water Research*, **45**(11), 3533–3540.

Bournazou M. N., Hooshiar K., Arellano-Garcia H., Wozny G. and Lyberatos G. (2013). Model based optimization of the intermittent aeration profile for SBRs under partial nitrification. *Water Research*, **47**(10), 3399–3410.

Cao Y. S. (2011). Mass Flow and Energy Efficiency of Municipal Wastewater Treatment Plants. IWA Publishing, London, UK.

Cao Y. and Pawłowski A. (2012). Sewage sludge-to-energy approaches based on anaerobic digestion and pyrolysis: brief overview and energy efficiency assessment. *Renewable and Sustainable Energy Reviews*, **16**(3), 1657–1665.

Cookney J., McLeod A., Mathioudakis V., Ncube P., Soares A., Jefferson B. and McAdam E. J. (2016). Dissolved methane recovery from anaerobic effluents using hollow fibre membrane contactors. *Journal of Membrane Science*, **502**, 141–150.

Crawford G. and Sandino J. (2010). Energy Efficiency in Wastewater Treatment in North America: A Compendium of best Practices and Case Studies of Novel Approaches. Water Environment Research Foundation/ IWA Publishing, Alexandria VA, USA/London UK.

Crone B. C., Garland J. L., Sorial G. A. and Vane L. M. (2016). Significance of dissolved methane in effluents of anaerobically treated low strength wastewater and potential for recovery as an energy product: a review. *Water Research*, **104**, 520–531.

Dai R., Liu Y., Liu X., Zhang X., Zeng C. and Li L. (2011). Investigation of a sewage-integrated technology combining an expanded granular sludge bed (EGSB) and an electrochemical reactor in a pilot-scale plant. *Journal of Hazardous Materials*, **192**(3), 1161–1170.

Diamantis V., Verstraete W., Eftaxias A., Bundervoet B., Siegfried V., Melidis P. and Aivasidis A. (2013). Sewage pre-concentration for maximum recovery and reuse at decentralized level. *Water Science and Technology*, **67**(6), 1188–1193.

Drechsel P., Qadir M. and Wichelns D. (2015). Wastewater: Economic Asset in an Urbanizing World. Springer, The Netherlands.

Florencio L., Kato M. T. and Cardoso de Morais J. (2001). Domestic sewage treatment in full-scale UASBB plant at Mangueira, Recife, Pernambuco. *Water Science and Technology*, **44**(4), 71–77.

Forster P., Ramaswamy V., Artaxo P., Berntsen T., Betts R., Fahey D. W., Haywood J., Lean J., Lowe D. C., Myhre G., Nganga J., Prinn R., Raga G., Schulz M. and Dorland R. V. (2007). Changes in atmospheric constituents and in radiative forcing. In: Climate Change 2007: The Physical Science Basis. Contribution of Working Group I to the Fourth Assessment Report of the Intergovernmental Panel on Climate Change, S. Solomon, D. Qin, M. Manning, Z. Chen, M. Marquis, K. B. Averyt, M. Tignor and H. L. Miller (eds). Cambridge University Press, United Kingdom and New York, NY, USA.

Ge S., Peng Y., Qiu S., Zhu A. and Ren N. (2014). Complete nitrogen removal from municipal wastewater via partial nitrification by appropriately alternating anoxic/aerobic conditions in a continuous plug-flow step feed process. *Water Researh*, **55**, 95–105.

Gilbert E. M., Agrawal S., Brunner F., Schwartz T., Horn H. and Lackner S. (2014). Response of different nitrospira species to anoxic periods depends on operational do. *Environ Science & Technology*, **48**(5), 2934–2941.

Goldstein R. and Smith W. (2002). Water & Sustainability (volume 4): US Electricity Consumption for Water Supply & Treatment-the Next Half Century. EPRI, Palo Alto, CA.

Gu J., Xu G. and Liu Y. (2017). An integrated AMBBR and IFAS-SBR process for municipal wastewater treatment towards enhanced energy recovery, reduced energy consumption and sludge production. *Water Research*, **110**, 262–269.

Gu J., Yang Q. and Liu Y. (2018a). Mainstream anammox in a novel A-2B process for energy-efficient municipal wastewater treatment with minimized sludge production. *Water Research*, **138**, 1–6.

Gu J., Yang Q. and Liu Y. (2018b). A novel strategy towards sustainable and stable nitritation-denitritation in an A-B process for mainstream municipal wastewater treatment. *Chemosphere*, **193**, 921–927.

Gu J., Zhang M., Wang S. Y. and Liu Y. (2019). Integrated upflow anaerobic fixed-bed and single-stage step-feed process for mainstream deammonification: A step further towards sustainable municipal wastewater reclamation. *Science of The Total Environment*, **678**, 559–564.

Guerrero J., Guisasola A. and Baeza J. A. (2011). The nature of the carbon source rules the competition between PAO and denitrifiers in systems for simultaneous biological nitrogen and phosphorus removal. *Water Research*, **45**(16), 4793–4802.

Gupta R. B., Basile A. and Veziroğlu T. N. (2015). Compendium of Hydrogen Energy: Hydrogen Storage, Distribution and Infrastructure. Woodhead Publishing, London, UK.

Harleman D. R., Susan M. and Chagnon F. (2002). Appropriate Wastewater Treatment in Developing Countries: Experiences with CEPT. See: https://www.pssurvival.com/PS/Water/Waste/Appropriate_Wastewater_Treatment_in_Developing_Countries_CEPT_2001.pdf (accessed 20 May 2019).

Harleman D. R. and Morrissey S. P. (1990). Chemically-Enhanced Treatment: An Alternative to Biological Secondary Treatment for Ocean Outfalls. Hydraulic Engineering. ASCE, 14–19.

Hellinga C., Schellen A., Mulder J. W., van Loosdrecht M. V. and Heijnen J. (1998). The SHARON process: an innovative method for nitrogen removal from ammonium-rich waste water. *Water Science and Technology*, **37**(9), 135–142.

Henderson M. (2002). Energy reduction methods in the aeration process at Perth wastewater treatment plant. University of Strathclyde, renewable energy systems and the environment MSc individual theses, (2001–2002).

Hendrickx T. L., Wang Y., Kampman C., Zeeman G., Temmink H. and Buisman C. J. (2012). Autotrophic nitrogen removal from low strength waste water at low temperature. *Water Research*, **46**(7), 2187–2193.

Jiang H., Liu G.-h., Ma Y., Xu X., Chen J., Yang Y., Liu X. and Wang H. (2018). A pilot-scale study on start-up and stable operation of mainstream partial nitrification-anammox biofilter process based on online pH-DO linkage control. *Chemical Engineering Journal*, **350**, 1035–1042.

Jimenez J., Miller M., Bott C., Murthy S., De Clippeleir H. and Wett B. (2015). High-rate activated sludge system for carbon management–Evaluation of crucial process mechanisms and design parameters. *Water Research*, **87**, 476–482.

Kim J., Kim K., Ye H., Lee E., Shin C., McCarty P. L. and Bae J. (2010). Anaerobic fluidized bed membrane bioreactor for wastewater treatment. *Environmental Science & Technology*, **45**(2), 576–581.

Kornaros M., Dokianakis S. N. and Lyberatos G. (2010). Partial nitrification/denitrification can be attributed to the slow response of nitrite oxidizing bacteria to periodic anoxic disturbances. *Environmental Science & Technology*, **44**(19), 7245–7253.

Kornaros M., Marazioti C. and Lyberatos G. (2008). A pilot scale study of a sequencing batch reactor treating municipal wastewater operated via the UP-PND process. *Water Science and Technology*, **58**(2), 435–438.

Kuenen J. G. (2008). Anammox bacteria: from discovery to application. *Nature Review Microbiology*, **6**, 320.

Lackner S., Welker S., Gilbert E. M. and Horn H. (2015). Influence of seasonal temperature fluctuations on two different partial nitritation-anammox reactors treating mainstream municipal wastewater. *Water Science and Technology*, **72**(8), 1358–1363.

Langone M., Ferrentino R., Cadonna M. and Andreottola G. (2016). Stoichiometric evaluation of partial nitritation, anammox and denitrification processes in a sequencing batch reactor and interpretation of online monitoring parameters. *Chemosphere*, **164**, 488–498.

Laureni M., Falas P., Robin O., Wick A., Weissbrodt D. G., Nielsen J. L., Ternes T. A., Morgenroth E. and Joss A. (2016). Mainstream partial nitritation and anammox: long-term process stability and effluent quality at low temperatures. *Water Research*, **101**, 628–639.

Laureni M., Weissbrodt D. G., Villez K., Robin O., De Jonge N., Rosenthal A., Wells G., Nielsen J. L., Morgenroth E. and Joss A. (2019). Biomass segregation between biofilm and flocs improves the control of nitrite-oxidizing bacteria in mainstream partial nitritation and anammox processes. *Water Research*, **154**, 104–116.

Leal C. D., Pereira A. D., Nunes F. T., Ferreira L. O., Coelho A. C., Bicalho S. K., Mac Conell E. F., Ribeiro T. B., de Lemos Chernicharo C. A. and de Araujo J. C. (2016). Anammox for nitrogen removal from anaerobically pre-treated municipal wastewater: effect of COD/N ratios on process performance and bacterial community structure. *Bioresource Technology*, **211**, 257–266.

Lemaire R., Marcelino M. and Yuan Z. G. (2008). Achieving the nitrite pathway using aeration phase length control and step-feed in an SBR removing nutrients from abattoir wastewater. *Biotechnology and Bioengineering*, **100**(6), 1228–1236.

Lemaire R., Zhao H., Thomson C., Christensson M., Piveteau S., Hemmingsen S., Veuillet F., Zozor P. and Ochoa J. (2014). Mainstream Deammonification with ANITATMMox Process. *Proceedings of the Water Environment Federation*, **10**(6), 2183–2197.

Lettinga G., Roersma R. and Grin P. (1983). Anaerobic treatment of raw domestic sewage at ambient temperatures using a granular bed UASB reactor. *Biotechnology and Bioengineering*, **25**(7), 1701–1723.

Li X., Xiao Y., Liao D., Zheng W., Yi T., Yang Q. and Zeng G. (2011). Granulation of simultaneous partial nitrification and anammox biomass in one single SBR system. *Applied Biochemistry and Biotechnology*, **163**(8), 1053–1065.

Liu Z. H., Yin H., Dang Z. and Liu Y. (2014). Dissolved methane: a hurdle for anaerobic treatment of municipal wastewater. *Environmental Science & Technology*, **48**(2), 889–890.

Liu Y.-J., Gu J. and Liu Y. (2018). Energy self-sufficient biological municipal wastewater reclamation: present status, challenges and solutions forward. *Bioresource Technology*, **269**, 513–519.

Lopez-Vazquez C. M., Oehmen A., Hooijmans C. M., Brdjanovic D., Gijzen H. J., Yuan Z. and van Loosdrecht M. C.M. (2009). Modeling the PAO–GAO competition: effects of carbon source, pH and temperature. *Water Research*, **43**(2), 450–462.

Lotti T., Kleerebezem R., van Erp Taalman Kip C., Hendrickx T. L., Kruit J., Hoekstra M. and van Loosdrecht M. C. (2014). Anammox growth on pretreated municipal wastewater. *Environmental Science & Technology*, **48**(14), 7874–7880.

Lotti T., Kleerebezem R., Hu Z., Kartal B., de Kreuk M. K., van Erp Taalman Kip C., Kruit J., Hendrickx T. L. and van Loosdrecht M. C. (2015). Pilot-scale evaluation of anammox-based mainstream nitrogen removal from municipal wastewater. *Environmental Technology*, **36**(9–12), 1167–1177.

Ma B., Zhang S., Zhang L., Yi P., Wang J., Wang S. and Peng Y. (2011). The feasibility of using a two-stage autotrophic nitrogen removal process to treat sewage. *Bioresource Technology*, **102**(17), 8331–8334.

Matsuura N., Hatamoto M., Sumino H., Syutsubo K., Yamaguchi T. and Ohashi A. (2010). Closed DHS system to prevent dissolved methane emissions as greenhouse gas in anaerobic wastewater treatment by its recovery and biological oxidation. *Water Science and Technology*, **61**(9), 2407–2415.

Matsuura N., Hatamoto M., Sumino H., Syutsubo K., Yamaguchi T. and Ohashi A. (2015). Recovery and biological oxidation of dissolved methane in effluent from UASB treatment of municipal sewage using a two-stage closed downflow hanging sponge system. *Journal of Environmental Management*, **151**, 200–209.

McCarty P. L., Bae J. and Kim J. (2011). Domestic wastewater treatment as a net energy producer: can this be achieved? *Environmental Science & Technology*, **45**(17), 7100–7106.

Meerburg F. A., Boon N., Van Winckel T., Vercamer J. A., Nopens I. and Vlaeminck S. E. (2015). Toward energy-neutral wastewater treatment: a high-rate contact stabilization process to maximally recover sewage organics. *Bioresource Technology*, **179**, 373–381.

Meerburg F. A., Boon N., Van Winckel T., Pauwels K. T. and Vlaeminck S. E. (2016). Live fast, die young: optimizing retention times in high-rate contact stabilization for maximal recovery of organics from wastewater. *Environmental Science & Technology*, **50**(17), 9781–9790.

Meinhold J., Arnold E. and Isaacs S. (1999). Effect of nitrite on anoxic phosphate uptake in biological phosphorus removal activated sludge. *Water Research*, **33**(8), 1871–1883.

Metcalf E. and Eddy M. (2014). Wastewater Engineering: Treatment and Resource Recovery. McGraw-Hill Education, New York.

Miller M. W., Bunce R., Regmi P., Hingley D. M., Kinnear D., Murthy S., Wett B. and Bott C. B. (2012). A/B process pilot optimized for nitrite shunt: high rate carbon removal followed by BNR with ammonia-based cyclic aeration control. *Proceedings of the Water Environment Federation*, **2012**(10), 5808–5825.

Miller M. W., Regmi P., Murthy S., Wett B. and Bott C. B. (2015). Combining high-rate activated sludge and shortcut nitrogen removal for efficient carbon and energy utilization. *Proceedings of the Water Environment Federation*, **2015**(2), 1–11.

Monteith H., Kalogo Y. and Louzeiro N. (2007). Achieving stringent effluent limits takes a lot of energy! *Proceedings of the Water Environment Federation*, **2007**(14), 4343–4356.

Morrissey S. P. and Harleman D. R.F. (1992). Retrofitting conventional primary treatment plants for chemically enhanced primary treatment in the USA. In: Chemical Water and Wastewater Treatment II: Proceedings of the 5th Gothenburg Symposium 1992, September 28–30, 1992, Nice, France, R. Klute and H. Hahn (eds). Springer Berlin Heidelberg, Berlin, Heidelberg, pp. 401–416.

Nges I. A. and Liu J. (2010). Effects of solid retention time on anaerobic digestion of dewatered-sewage sludge in mesophilic and thermophilic conditions. *Renewable Energy*, **35**(10), 2200–2206.

Nogaj T. M., Randall A. A., Jimenez J. A., Takacs I., Bott C. B., Miller M. W., Murthy S. and Wett B. (2014). Mathematical modeling of the high rate activated sludge system: optimizing the COD: N ratio in the process effluent. *Proceedings of the Water Environment Federation*, 2014(16), 913–2926.

Parker D. S., Barnard J., Daigger G. T., Tekippe R. J. and Wahlberg E. J. (2001). The future of chemically enhanced primary treatment: evolution not revolution. *WATER*, **21**, 49–56.

Parkin G. F. and Owen W. F. (1986). Fundamentals of anaerobic digestion of wastewater sludges. *Journal of Environmental Engineering*, **112**(5), 867–920.

Rahman A., De Clippeleir H., Thomas W., Jimenez J. A., Wett B., Al-Omari A., Murthy S., Riffat R. and Bott C. (2019). A-stage and high-rate contact-stabilization performance comparison for carbon and nutrient redirection from high-strength municipal wastewater. *Chemical Engineering Journal*, **357**, 737–749.

Regmi P., Miller M. W., Holgate B., Bunce R., Park H., Chandran K., Wett B., Murthy S. and Bott C. B. (2014). Control of aeration, aerobic SRT and COD input for mainstream nitritation/denitritation. *Water Research*, **57**, 162–171.

Regmi P., Holgate B., Fredericks D., Miller M. W., Wett B., Murthy S. and Bott C. B. (2015). Optimization of a mainstream nitritation-denitritation process and anammox polishing. *Water Science and Technology*, **72**(4), 632–642.

Regmi P., Holgate B., Miller M. W., Park H., Chandran K., Wett B., Murthy S. and Bott C. (2016). Nitrogen polishing in a fully anoxic anammox MBBR treating mainstream nitritation–denitritation effluent. *Biotechnology and Bioengineering*, **113**(3), 635–642.

Rossle W. and Pretorius W. (2001). A review of characterisation requirements for in-line prefermenters: paper 1: wastewater characterisation. *Water Sa*, **27**(3), 405–412.

Semerjian L. and Ayoub G. M. (2003). High-pH–magnesium coagulation–flocculation in wastewater treatment. *Advances in Environmental Research*, **7**(2), 389–403.

Smith A. L., Skerlos S. J. and Raskin L. (2013). Psychrophilic anaerobic membrane bioreactor treatment of domestic wastewater. *Water Research*, **47**(4), 1655–1665.

Smitshuijzen J., Perez J., Duin O. and van Loosdrecht M. C. M. (2016). A simple model to describe the performance of highly-loaded aerobic COD removal reactors. *Biochemical Engineering Journal*, **112**, 94–102.

Strous M., Heijnen J. J., Kuenen J. G. and Jetten M. S. M. (1998). The sequencing batch reactor as a powerful tool for the study of slowly growing anaerobic ammonium-oxidizing microorganisms. *Applied Microbiology and Biotechnology*, **50**(5), 589–596.

Strous M., Kuenen J. G. and Jetten M. S. (1999). Key physiology of anaerobic ammonium oxidation. *Applied and. Environmental Microbiology*, **65**(7), 3248–3250.

Sun F., Su X., Kang T., Wu S., Yuan M., Zhu J., Zhang X., Xu F. and Wu W. (2016). Integrating landfill bioreactors, partial nitritation and anammox process for methane recovery and nitrogen removal from leachate. *Scientific Reports*, **6**, 27744.

van der Star W. R., Abma W. R., Blommers D., Mulder J. W., Tokutomi T., Strous M., Picioreanu C. and van Loosdrecht M. C. (2007). Startup of reactors for anoxic ammonium oxidation: experiences from the first full-scale anammox reactor in Rotterdam. *Water Research*, **41**(18), 4149–4163.

van Kempen R., Mulder J. W., Uijeterlinde C. A. and Loosdrecht M. C. M. (2001). Overview: full scale experience of the SHARON® process for treatment of rejection water of digested sludge dewatering. *Water Science and Technology*, **44**(1), 145–152.

Wan J., Gu J., Zhao Q. and Liu Y. (2016). COD capture: a feasible option towards energy self-sufficient domestic wastewater treatment. *Science Report*, **6**, 25054.

Wang Q. (2017). A roadmap for achieving energy-positive sewage treatment based on sludge treatment using free ammonia. *ACS Sustainable Chemistry & Engineering*, **5**(11), 9630–9633.

Wang H., Li F., Keller A. A. and Xu R. (2009). Chemically enhanced primary treatment (CEPT) for removal of carbon and nutrients from municipal wastewater treatment plants: a case study of Shanghai. *Water Science and Technology*, **60**(7), 1803–1809.

Wang Q., Ye L., Jiang G., Hu S. and Yuan Z. (2014). Side-stream sludge treatment using free nitrous acid selectively eliminates nitrite oxidizing bacteria and achieves the nitrite pathway. *Water Research*, **55**, 245–255.

Wang G., Xu X., Zhou L., Wang C. and Yang F. (2017a). A pilot-scale study on the start-up of partial nitrification-anammox process for anaerobic sludge digester liquor treatment. *Bioresource Technology*, **241**, 181–189.

Wang Q., Duan H., Wei W., Ni B.-J., Laloo A. and Yuan Z. (2017b). Achieving Stable Mainstream Nitrogen Removal via the Nitrite Pathway by Sludge Treatment Using Free Ammonia. *Environmental Science & Technology*, **51**(17), 9800–9807.

Wang C., Liu S., Xu X., Zhang C., Wang D. and Yang F. (2018). Achieving mainstream nitrogen removal through simultaneous partial nitrification, anammox and denitrification process in an integrated fixed film activated sludge reactor. *Chemosphere*, **203**, 457–466.

Wang H., Xu G., Qiu Z., Zhou Y. and Liu Y. (2019). NOB suppression in pilot-scale mainstream nitritation-denitritation system coupled with MBR for municipal wastewater treatment. *Chemosphere*, **216**, 633–639.

Wett B., Buchauer K. and Fimml C. (2007). Energy self-sufficiency as a feasible concept for wastewater treatment systems. In: IWA Leading Edge Technology Conference. Singapore, pp. 21–24.

Wett B., Omari A., Podmirseg S. M., Han M., Akintayo O., Gomez Brandon M., Murthy S., Bott C., Hell M., Takacs I., Nyhuis G. and O'Shaughnessy M. (2013). Going for mainstream deammonification from bench to full scale for maximized resource efficiency. *Water Science and Technology*, **68**(2), 283–289.

Whang L. M., Chien I. C., Yuan S. L. and Wu Y. J. (2009). Nitrifying community structures and nitrification performance of full-scale municipal and swine wastewater treatment plants. *Chemosphere*, **75**(2), 234–242.

Wyffels S., Boeckx P., Pynaert K., Zhang D., Van Cleemput O., Chen G. and Verstraete W. (2004). Nitrogen removal from sludge reject water by a two-stage oxygen-limited autotrophic nitrification denitrification process. *Water Science and Technology*, **49** (5–6), 57–64.

Xu G., Zhang Y. and Gregory J. (2006). Different pollutants removal efficiencies and pollutants distribution with particle size of wastewater treated by CEPT process. *Water Practice and Technology*, **1**(3), wpt2006047.

Xu G., Zhou Y., Yang Q., Lee Z. M., Gu J., Lay W., Cao Y. and Liu Y. (2015). The challenges of mainstream deammonification process for municipal used water treatment. *Applied Microbiology and Biotechnology*, **99**(6), 2485–2490.

Xu G., Wang H., Gu J., Shen N., Qiu Z., Zhou Y. and Liu Y. (2017). A novel A-B process for enhanced biological nutrient removal in municipal wastewater reclamation. *Chemosphere*, **189**, 39–45.

Xu X., Wang G., Zhou L., Yu H. and Yang F. (2018). Start-up of a full-scale SNAD-MBBR process for treating sludge digester liquor. *Chemical Engineering Journal*, **343**, 477–483.

Yamamoto S., Alcauskas J. B. and Crozier T. E. (1976). Solubility of methane in distilled water and seawater. *Journal of Chemical and Engineering Data*, **21**(1), 78–80.

Yang S. and Yang F. (2011). Nitrogen removal via short-cut simultaneous nitrification and denitrification in an intermittently aerated moving bed membrane bioreactor. *Journal of Hazardous Materials*, **195**, 318–323.

Zheng Z., Huang S., Bian W., Liang D., Wang X., Zhang K., Ma X. and Li J. (2019). Enhanced nitrogen removal of the simultaneous partial nitrification, anammox and denitrification (SNAD) biofilm reactor for treating mainstream wastewater under low dissolved oxygen (DO) concentration. *Bioresource Technology*, **283**, 213–220.

Zhou Y., Zhang D. Q., Le M. T., Puah A. N. and Ng W. J. (2013). Energy utilization in sewage treatment – a review with comparisons. *Journal of Water and Climate Change*, **4**(1), 1.

Chapter 4

Emerging biological processes for municipal wastewater treatment

Meng Zhang, Hang Liu, Bin Ji and Yu Liu

4.1 INTRODUCTION

Today, a paradigm shift of biological processes is necessary to move towards energy self-sufficient municipal wastewater treatment. The A-B process configurations presented in Chapter 3 provide a feasible engineering solution, in which the A-stage is primarily designed for chemical oxygen demand (COD) capture for energy recovery instead of biological oxidation, while the B-stage is designated for more cost-effective nutrient removal. It is obvious that the central point of the A-B process is to separate the carbon and nitrogen removal processes, with a reduced demand on COD for nitrogen removal. Process analysis (Chapter 2) also shows that the A-B process with anammox as a core appears to be a highly feasible option for concurrently realizing energy recovery from wastewater and energy savings from nitrogen removal. Although the anammox process has been well-developed for treating ammonia-rich wastewater (e.g. anaerobic digester liquor) with lots of successful large-scale applications worldwide (Lackner *et al.*, 2014), mainstream deammonification still remains challenging. This chapter thus presents two mainstream approaches, i.e. partial nitrification and partial denitrification-anammox, whilst the associated challenges and possible solutions are also discussed.

Water shortage is now becoming a global issue, potentially threatening human health. It has been reported that the current water supply will only satisfy about

Author(s): Yu Liu, Jun Gu and Meng Zhang
doi: 10.2166/9781789060089_0113

60% of the global water demand by 2030 (Boccaletti *et al.*, 2010). Similarly, resource scarcity is also now at a critical moment for society and current levels of use are not sustainable. For example, it is estimated that phosphorus will be exhausted in the next 50–400 years in the scenario of business as usual (Reijnders, 2014). As such, municipal wastewater should be viewed as a source rich in water and various resources. To realize synergetic recovery of water, energy and resources from municipal wastewater, two emerging biological technologies, namely the integrated anaerobic membrane bioreactor (AnMBR) and reverse osmosis (RO) process and the microalgal-bacterial process are investigated in this chapter, given their high potentials for concurrent water reclamation and resource recovery.

4.2 MAINSTREAM DEAMMONIFICATION PROCESSES

Mainstream deammonification has been regarded as a viable technology to achieve energy self-sufficient municipal wastewater treatment due to its significant reductions in oxygen and COD demand with less biosludge production. Anaerobic ammonia oxidation, known as anammox, is the core process of mainstream deammonification, where anammox bacteria utilize nitrite (NO_2^-) and ammonium (NH_4^+) as substrates to produce nitrogen gas (N_2) without consuming COD (Kartal *et al.*, 2010). Due to the fact that ammonium already exists in municipal wastewater, a successful anammox process is indeed largely dependent on a sustainable supply of nitrite. So far, partial nitrification (PN) and partial denitrification (PDN) are the two popular approaches to serve this purpose (Ahn *et al.*, 2008; Kalyuzhnyi *et al.*, 2006), leading to the development of PN-anammox and PDN-anammox for treating municipal wastewater.

4.2.1 Approach I: PN-anammox

In the PN-anammox process, about 57% of influent ammonia is supposed to be oxidized to nitrite via PN, and the produced nitrite further reacts with the remaining influent ammonia via anammox to produce nitrogen gas as the major end product. According to Equation 4.1, in the anammox reaction (Strous *et al.*, 1998), theoretically, about 89% of wastewater ammonia is removable through PN-anammox:

$$\begin{aligned} &NH_4^+ + 1.32NO_2^- + 0.066HCO_3^- + 0.13H^+ \rightarrow 1.02N_2 + 0.26NO_3^- \\ &\quad + 0.066CH_2O_{0.5}N_{0.15} + 2.03H_2O \end{aligned} \tag{4.1}$$

It should be noted that PN-anammox is a complete autotrophic process without the needs of organics for nitrogen removal. As such, municipal wastewater COD can be pre-captured as much as possible, benefiting the maximized energy recovery and minimized waste sludge production. This in turn provides a unique opportunity to achieve energy neutral municipal wastewater treatment. Because of the remarkable advantages, mainstream PN-anammox had been proposed to treat municipal wastewater since the late 1990s (Jetten *et al.*, 1997). So far, mainstream PN-anammox has been successfully performed in laboratory biofilm and granular sludge reactors fed with synthetic or real municipal wastewater with high total nitrogen removal of 70–90% (Cao *et al.*, 2017c). For example, PN-anammox can be realized in single-stage and two-stage processes with high rate activated sludge (HRAS) and anaerobic treatment as COD pre-capture technologies, as illustrated in Chapter 3.

Pilot-scale PN-anammox processes have been developed to treat municipal wastewaters, which can be realized in single- and two-stage configurations (Table 4.1). In the single-stage process, both PN and anammox are realized in one reactor, e.g. Lotti *et al.* (2015a) reported a single-stage PN-anammox reactor of 4 m^3 with granular sludge, which was fed with wastewater of low C/N ratio. In this configuration, 50–70% of total nitrogen removal was achieved, with a nitrogen removal rate of 0.2–0.3 kg $N/(m^3 \cdot d)$ at a temperature of 17–23°C. Moving bed biofilm reactors (MBBRs) and their hybrid systems have also been explored for PN-anammox (Gustavsson *et al.*, 2014; Trojanowicz *et al.*, 2016; Vilpanen, 2017). In a two-stage configuration, the PN and anammox reactions are realized in two separate reactors, with the first reactor being suspended sludge, and the second one being biofilms or granular sludge, enabling PN and anammox to be optimized separately. For example, the anammox reactor can be operated in an anoxic environment to avoid the competition of nitrite oxidizing bacteria (NOB) on nitrite. A two-stage pilot-scale PN-anammox was successfully performed with high total removal of about 75% at 23°C (Regmi *et al.*, 2015). Compared with the two-stage system, single-stage PN-anammox has a simple process configuration with reduced capital and operation costs, and less emission of nitric and nitrous oxide gas (Massara *et al.*, 2017; Zhang *et al.*, 2019a). In practice, single-stage PN-anammox appears more popular in mainstream deammonification applications. Additionally, it should be noted that the performance of pilot-scale PN-anammox (Table 4.1) is comparable with or even better than that (0.05–0.1 kg $N/(m^3 \cdot d)$) achieved in the conventional nitrification and denitrification process (Henze *et al.*, 2008), showing the great potential of PN-anammox in mainstream nitrogen removal.

So far, two full-scale mainstream PN-anammox have been reported: in the Strass Wastewater Treatment Plant (WWTP), Austria, and Changi Water Reclamation Plant (WRP), Singapore. To sustain deammonification, three operation strategies were adopted in the Strass WWTP, i.e. selection of anammox granules in the mainstream by cyclone, bio-augmentation using anammox sludge and ammonia oxidizing bacteria (AOB) harvested from a sidestream deammonification process, and intermittent aeration to repress nitrite oxidizers (Figure 4.1) (Wett *et al.*, 2013). In the Changi WRP, PN-anammox was reported in one train (200,000 m^3/d) of the five-pass step-feed activated sludge process, likely due to the short sludge retention time and tropical climate with an annual temperature of 28–32°C

Table 4.1 Examples of the mainstream PN-anammox process at pilot- and full-scale.

Process/Biomass	Reactor Volume, m^3	HRT, h	T, °C	Influent NH_4^+-N, mg/L	Influent COD/N	N Removal Rate, kg N/(m^3·d)	TN Removal, %	References
Pilot-scale								
Single-stage activated sludge	0.2	3.7	Ambient	24	1–3	0.1	70	Han *et al.* (2016)
Single-stage MBBR (K1 carriers)	0.2	38–40	15–17	45	1.0–1.6	0.05–0.10	25–51	Trojanowicz *et al.* (2016)
Single-stage MBBR (K1 carriers)	2.3	4.0	15	28	2.0	0.12	60	Gustavsson *et al.* (2014)
Single-stage IFAS MBBR reactor (K5 carriers) ANITA™ Mox process	3.0	7.1	≥15	36	–	0.09	65	Vilpanen (2017)
Single-stage granular sludge	4.0	1.5–2.0	17–23	27	0.67	0.2–0.3	50–70	Lotti *et al.* (2015a)
Two-stage PN (activated sludge)	0.34	2–6	23	33	8.9	0.12[a]	60	Regmi *et al.* (2015)
Two-stage Anammox (MBBR)	0.45	1.1–3.4		6.6 (NH_4^+-N) 2.1 (NO_2^--N) 4.4 (NO_3^--N)	3.9	0.05[a]	38	
Full-scale								
Single-stage activated sludge with argumentation	10,460	5.6	9–15	15–40	9.2	0.15	82	Wett *et al.* (2013)
Step-feed activated sludge	56,800	5.6	30	30.8	~7	0.15	88	Cao *et al.* (2013)

IFAS = integrated fixed-film activated sludge; MBBR = moving bed biofilm reactor; TN = total nitrogen.
[a]Calculated according to influent nitrogen concentrations, hydraulic retention time and nitrogen removal efficiency.

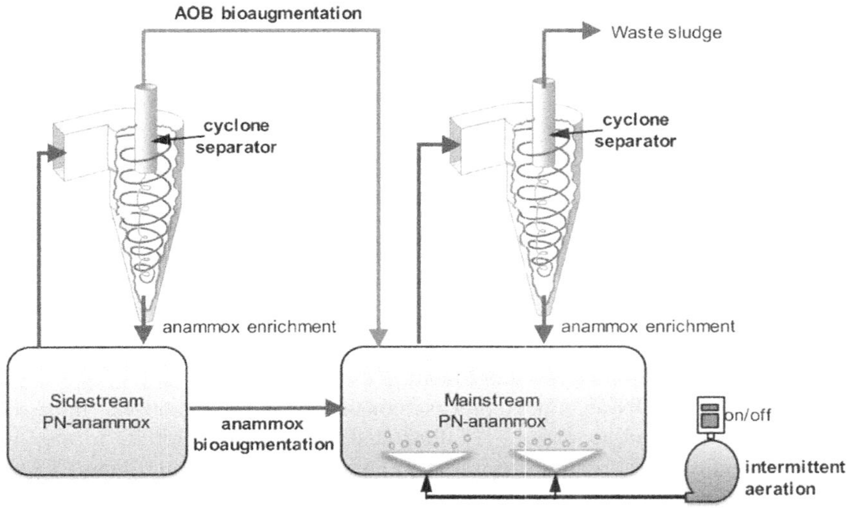

Figure 4.1 Illustration of the operation strategies for sustaining mainstream PN-anammox in the Strass WWTP. Adapted from Wett *et al.* (2013), with permission.

(Cao *et al.*, 2013). However, it should also be realized that the observed nitrogen removal was likely enhanced by conventional heterotrophic denitrification in consideration of the high influent C/N ratio. For example, conventional denitrification contributed to 44.6% of the total nitrogen removal in the Changi WRP (Cao *et al.*, 2017b). Obviously, more work is needed to convincingly demonstrate the feasibility and long-term stability of deammonification for treating municipal wastewater.

4.2.2 Approach II: PDN-anammox

The mainstream deammonification process can also be realized through partial denitrification (PDN)-anammox. In the PDN-anammox process, nitrite is produced through the partial denitrification of nitrate, which can be achieved via autotrophic or heterotrophic denitrification. Autotrophic PDN uses inorganic compounds as the electron donors, e.g. sulphide may be an electron donor for DEnitrifying AMmonium OXidation (DEAMOX) and iron(II) for Nitrate-dependent Anaerobic Ferrous Oxidation (NAFO) (Kalyuzhnyi *et al.*, 2006; Zhang *et al.*, 2015). By contrast, heterotrophic PDN makes use of organic carbon as the electron donor. Given the relatively high COD concentration in municipal wastewater, heterotrophic PDN is dominant over autotrophic PDN in practical situations. Thus, this section primarily concentrates on heterotrophic PDN-anammox. In the heterotrophic PDN process, part influent COD is required for converting nitrate to nitrite, meaning that PDN-anammox indeed is a hybrid heterotrophic-autotrophic

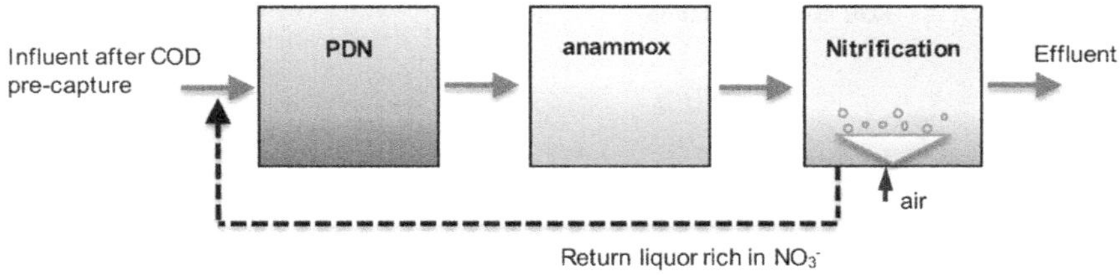

Figure 4.2 Schematic illustration of a PDN-anammox process. Inspired by Ma *et al.* (2017).

process, and pre-capture of COD is needed in order to create a favourable COD/N ratio (typically 2.0–3.0) for the subsequent PDN-anammox.

Generally, the PDN-anammox process consists of three operational units, i.e. nitrification, PDN and anammox. As illustrated in Figure 4.2, nitrate formed via nitrification is partially reduced to nitrite in an anoxic tank with influent COD as the electron donor. The generated nitrite then reacts with part of the influent ammonia in the anammox unit, while the remaining ammonia is oxidized to nitrate via complete nitrification. According to Equations 4.1, 4.2 and 4.3, about 53% of influent ammonia should be oxidized to nitrate through complete nitrification, and wastewater ammonia is removable through PDN-anammox (Zhang *et al.*, 2019b):

$$NH_4^+ + 1.82O_2 + 0.13CO_2 \rightarrow 0.03C_5H_7NO_2 + 0.97NO_3^- + 0.92H_2O + 1.97H^+ \quad (4.2)$$

$$\begin{aligned} NO_3^- + 0.37CH_3COO^- + 0.036H^+ \rightarrow\ & 0.036C_5H_7NO_2 + 0.97NO_2^- \\ & + 0.37HCO_3^- + 0.26H_2O + 0.19CO_2 \end{aligned} \quad (4.3)$$

However, compared to PN-anammox, an additional line is often required to return nitrification liquor rich in nitrate back to the pre-PDN unit for partial denitrification.

In the PDN process, nitrite is accumulated for the subsequent anammox reaction through partial denitrification in which nitrite production significantly depends on the differences between the nitrate reduction rate (NaRR) and the nitrite reduction rate (NiRR). As can be seen in Table 4.2, the NaRR/NiRR ratio ranges widely from 1.1 to 16.9, indicating various capabilities of nitrite accumulation through PDN. In fact, a high nitrite accumulation rate (NAR) of over 90% was achievable in laboratory systems fed with synthetic or real municipal wastewater with the low C/N ratios. In addition, an anoxic/oxic biofilm-based system was employed to treat synthetic municipal wastewater via PDN-anammox, with a total nitrogen removal efficiency of 80 $\pm$ 4% at a low influent C/N ratio of 2.6, of which more than 70% of nitrogen was removed through anammox in the anoxic reactor (Ma *et al.*, 2017). Du *et al.* (2019b) introduced a high-strength nitrate wastewater to

Table 4.2 Nitrite production potential via PDN in laboratory-scale wastewater treatment processes.

Nitrate, mg N/L	NaRR, mg N/(g VSS h)	NiRR, mg N/(g VSS h)	NaRR/NiRR	NAR, %	C/N ratio	References
20–150	59.61–82.31	7.30–9.35	8.17–9.36	~89	3.0	Du *et al.* (2016a)
60.5	56.61–82.31	7.30–8.80	7.6–9.4	80	2.6	Ma *et al.* (2017)
60	7.21	5.92	1.2	84	2.0–3.0	Si *et al.* (2018)
60	24.0	1.7	14.1	90	3.0	Ji *et al.* (2018)
60	50.6	48.5	1.04	>90	3.4	Du *et al.* (2019a)
59.3	190	164.17	1.16	90	2.5	Cao *et al.* (2017c)
50	4.5–19.53	2.8–17.69	1.1–1.6	95.8	3.0	Du *et al.* (2017)
50	0.9–13.3	0.5–8.4	1.5–1.8	87	3.0	Du *et al.* (2017)
40	82.2	74.4	1.1	80	3.0	Du *et al.* (2016b)

NaRR = nitrate reduction rate; NiRR = nitrite reduction rate; NAR = nitrite accumulation rate, i.e. the ratio of produced nitrite to influent nitrate in PDN process; VSS = volatile suspended solids.

enhance partial denitrification in a PDN-anammox process, i.e. a small volume of high-strength nitrate wastewater was pumped into a PDN reactor to promote nitrite production, and partially denitrified liquor was then blended with a large volume of low-strength municipal wastewater for the oxidation of ammonia via anammox.

The research into PDN-anammox as an alternative for nitrogen removal is still in an early development stage, and there is a lack of demonstrations of PDN-anammox at pilot- and full-scale. However, it should be realized that the process configuration of PDN-anammox as described in Figure 4.2 is, to some extent, similar to the anaerobic-anoxic-oxic (A^2O) process that has been adopted in full-scale WWTPs (Jin *et al.*, 2014). This in turn implies that there is a possibility to incorporate PDN-anammox into the current A^2O process. For example, PDN-anammox was recently reported to enhance nitrogen removal in a large-scale anaerobic-anoxic-oxic process for municipal wastewater treatment where an anoxic tank with suspended culture was converted to a moving-bed bioreactor (Li *et al.*, 2019).

4.2.3 Challenges of mainstream deammonification

The deammonification processes have exhibited numerous advantages, as discussed above. However, deammonification has only been successfully applied for treating ammonium-rich wastewater, e.g. anaerobic digester liquor, landfill leachate etc. (Fux *et al.*, 2006; Kulikowska & Bernat, 2013; Lackner *et al.*, 2014). Although some successful cases at the lab- and pilot-scale have been reported, it appears from the current literature that the up-scaling of mainstream deammonification for municipal wastewater treatment still remains challenging. Thus, some technical issues related to process feasibility and stability are elucidated in this section.

4.2.3.1 Stable nitrite supply

In short-cut biological nitrogen removal processes, ammonia is primarily removed via nitrite. Thus, a constant and stable nitrite supply is a prerequisite for mainstream short-cut biological nitrogen removal. As shown in Figure 4.3, nitrite can be produced in both nitrification and denitrification via the mediation of different functional species. It appears that the activity and growth of nitrite oxidizing bacteria (NOB) should be suppressed against ammonia oxidizing bacteria (AOB) for stable nitrite accumulation in the PN-anammox process, while nitrate reduction should be stopped at the nitrite stage in PDN-anammox process.

In general, the specific growth rate of AOB is about 1.5–2.0 times higher than that of NOB (Rittmann & McCarty, 2001). This may offer the great possibility of selectively enriching AOB over NOB. So far, several operation strategies have been reported for suppression of NOB while retaining AOB, such as dissolved

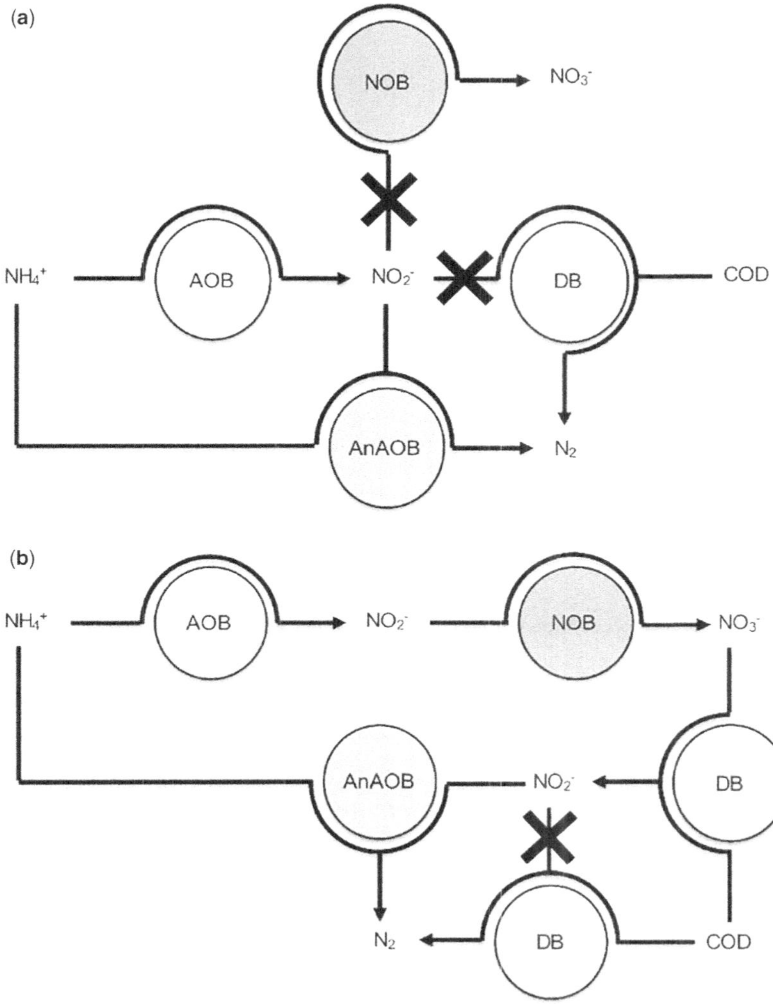

Figure 4.3 Challenges associated with (a) PN-anammox and (b) PDN-anammox. AOB: ammonia oxidizing bacteria; NOB: nitrite oxidizing bacteria; AnAOB: anaerobic ammonia oxidizing bacteria; DB: denitrifying bacteria. Adapted from Zhang *et al.* (2019b), with permission.

oxygen (DO), sludge retention time (SRT), free ammonia (FA) and free nitrous acid (FNA), controlled aeration etc (Figure 4.4):

- **DO:** the half-saturation coefficient of DO was found to be 0.2–0.4 mg/L for AOB, which was much lower than that for NOB falling into the range

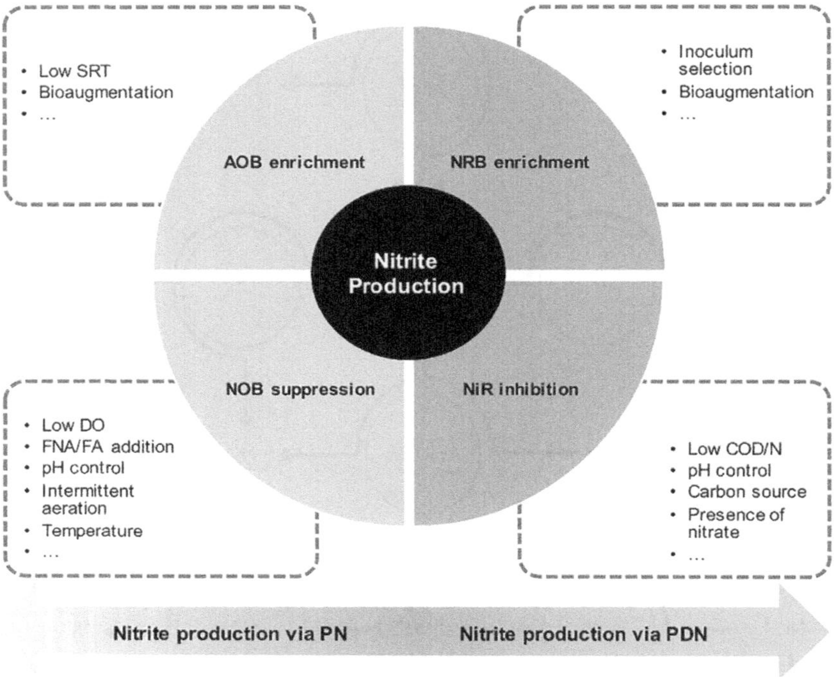

Figure 4.4 Control strategies for nitrite production via PN and PDN processes. Zhang *et al.* (2019b), with permission.

of 1.2–1.5 mg/L (Picioreanu *et al.*, 1997), i.e. a low DO concentration favours the selection of AOB. In another example, the growth rate of AOB was about 3–4 times higher than that of NOB under DO-limiting conditions (Ahn *et al.*, 2008). However, it should be noted that low DO concentration may also comprise AOB activity and trigger filamentous growth (Guo *et al.*, 2010). On the other hand, the NOB-oxygen affinity varies among different genera of NOB. For example, *Nitrobacter can* well adapt to high DO concentration, while *Nitrospira*-like nitrite oxidizers grew well at a low DO concentration of less than 0.5 mg/L (Daims *et al.*, 2001; Regmi *et al.*, 2014). Generally, DO concentration may be controlled at about 1.5 mg/L or even lower for NOB suppression, but it should not negatively impact on AOB activity (Ruiz *et al.*, 2006).

- **SRT:** AOB present a higher growth rate than that of NOB under certain conditions, such as low DO concentration and high temperature. This suggests that AOB can be selectively retained or enriched in the system by controlling SRT at a level longer than the doubling time (the time taken by the bacteria to double in number) of AOB, while shorter than that of NOB

(Xu *et al.*, 2015). For example, effective retention of AOB against NOB was achieved at a temperature of 30–40°C and SRT of 1.5–2.0 days in a nitritation-denitritation process (Hellinga *et al., 1998*).

- **Alternate oxic-anoxic cycle:** in general, AOB has a shorter lag time than NOB when transiting between anoxic to oxic conditions. Such an alternate transition might trigger the deactivation of a key enzyme that inhibits NOB directly or causes its inhibition by intermediate products such as hydroxylamine (Bournazou *et al.*, 2013; Xu *et al.*, 2012). Therefore, an intermittent aeration strategy was proposed for suppression of NOB activity by creating transient anoxic conditions in an SBR or a step-feed reactor (Gu *et al.*, 2018; Wett *et al.*, 2013). Additionally, to enhance nitrite accumulation, real-time control was also proposed for regulating the lengths of the oxic and anoxic phases based on in-line sensing, e.g. oxidation-reduction potential, oxygen uptake rate, etc. (Regmi *et al.*, 2014).
- **FA and FNA inhibition:** the inhibition of FA and FNA on nitrification was first reported by Anthonisen *et al.* (1976). Evidence shows that NOB is more sensitive to both FA and FNA than AOB. This may provide an opportunity for controlling nitrite accumulation in wastewater with high ammonia concentration (Park *et al.*, 2010). However, it should be realized that such a control strategy is no longer effective in mainstream municipal wastewater treatment with low ammonia concentrations of below 100 mg/L (Henze *et al.*, 2008). To tackle such a challenge, FA and FNA from the sidestream digester liquor has been proposed to be collected and recirculated to the mainstream treatment process (Liu *et al.*, 2019).

In practice, a single parameter-based control strategy was demonstrated to be unable to sustain the effective suppression of NOB against AOB. In this regard, Wang *et al.* (2019) reported a holistic operation strategy by synergetic control of DO, SRT and sludge return ratio. As a result, successful suppression of NOB was achieved with a nitrite accumulation rate of more than 80% in a pilot step-feed plant with a treatment capacity of 30 m^3 real municipal wastewater per day.

Conventional denitrification consists of four reductive reactions, i.e. $NO_3^- \rightarrow NO_2^- \rightarrow NO \rightarrow N_2O \rightarrow N_2$ (Shapleigh, 2013). To sustain the nitrite supply via PDN, nitrate reduction should be controlled at the nitrite step. Currently, two approaches have been explored for this purpose: (i) enriching specific bacteria that can only convert nitrate to nitrite, i.e. nitrate respiring bacteria (NRB). NRB (e.g. *Pseudomonas*, *Citrobacter*, *Terrabacter* and *Thauera*) have the capability of accumulating nitrite due to the lack of nitrite reductase (NiR) (Glass & Silverstein 1998, Liu *et al.*, 2013a, Roco *et al.*, 2017). However, it appears difficult to maintain the long-term dominance of these NRB, especially in large-scale engineering settings; (ii) suppressing the activity of NiR against nitrate reductase, leading to an enhanced nitrate reduction rate, while a reduced nitrite reduction rate. Due to the fact that cytochrome *c* or the quinol pool in the respiratory chain is the electron source for both nitrate and nitrite reductases (Chen & Strous, 2013), a low COD/N ratio of 2.0–3.0 generally may lead to a situation lacking electrons for further reduction of nitrite to nitric oxide (Table 4.2). As a result, nitrite accumulation is to be expected (Cao *et al.*, 2017a; Du *et al.*, 2016a; Gong *et al.*, 2013). Moreover, other operational parameters, such as carbon source, hydraulic retention time (HRT), pH, nitrate etc., have all been examined for regulating nitrite accumulation (Figure 4.4).

4.2.3.2 Proper COD/N ratio

Raw municipal wastewater has relatively high COD while having low ammonia concentrations at the mass ratio of 10–14, which decrease to 7–10 after primary settling (Henze *et al.*, 2008; Metcalf & Eddy, 2014). However, this COD/N ratio is still higher than that required for deammonification. As illustrated in Figure 4.3a, in PN-anammox, heterotrophic denitrifying bacteria can compete with autotrophic anammox bacteria for nitrite when COD is available. This in turn seriously compromises the accumulation of nitrite. Due to unfavourable thermodynamics, it appears to be extremely difficult to maintain a substantial amount of slow-growing anammox bacteria in a condition where COD is sufficient (Jenni *et al.*, 2014; Kumar & Lin, 2010). A biodegradable COD/N ratio of below 0.5 has been reported to maintain a sustainable deammonification (Daigger, 2014). The situation in PDN-anammox is even more complicated. For example, COD should be first eliminated partially from the influent wastewater to avoid possible competition between anammox and denitrifying bacteria, while enough COD should remain for the subsequent heterotrophic PDN process by which nitrate produced through nitrification is converted to nitrite, blocking further reduction to nitrogen gas (Figure 4.3b). Table 4.2 showed that a COD/N ratio of 2.0–3.0 might be favourable for PDN. This clearly suggests that prior capture of influent COD is needed for successful mainstream deammonification processes. Considering the low COD/N ratio required for PN-anammox and PDN-anammox, efficient COD capture technologies should be integrated into any mainstream deammonification process, e.g. with an HRAS and anaerobic unit.

4.2.3.3 Temperature

Temperature is a crucial parameter for mainstream deammonification and can affect the metabolic activity, cell growth and community structure of microorganisms. Anammox bacteria, as the core functional microbes of mainstream deammonification, are very sensitive to temperature. For example, it has been reported that the specific activity of anammox bacteria could decrease approximately 10 times when the temperature drops from approximately 30°C to 10°C, with a sharper activity decline being observed at 15°C (Lotti *et al.*, 2015b). The observed decline in anammox activity was likely caused by reactive oxygen species or toxic nitrogen compounds in anammox bacteria at the lower temperatures (Lin *et al.*, 2018). On the other hand, temperature has a stronger effect on anammox bacteria than on AOB and denitrifiers. On average, the maximum specific growth rate of anammox bacteria was found to be about 0.05–0.09/day at 30°C (Strous *et al.*, 1999), which was 10 times lower than that of AOB and denitrifiers (Cao *et al.*, 2017c; Rittmann and McCarty, 2001). As such, it is difficult to create an environment where the population and activities of key groups are well balanced towards biological nitrogen removal. It should also be realized that wastewater temperature may vary in a large range due to seasonal changes and location diversity. Obviously, lower temperature encountered in winter would adversely affect deammonification, while challenging the large-scale applications of mainstream deammonification. It has been suggested that anammox biofilms or granules should be cultivated in an aerated PN-anammox process to mitigate the adverse effects of low temperature (Lotti *et al.*, 2015b). Some laboratory and pilot studies have shown successful PN-anammox at temperatures of 15–25°C, but with a long adaptation period.

4.3 INTEGRATED ANAEROBIC MEMBRANE BIOREACTOR AND REVERSE OSMOSIS PROCESS

With the rapid growth of population and developing urbanization, water scarcity is becoming increasingly severe. To address such a global water issue, treated municipal wastewater should be viewed as a potential water resource. Extensive efforts have thus been dedicated to producing water with appropriate quality for water recycling and reuse. For example, in Singapore, high-grade NEWater is reclaimed from secondary effluent and has become an important water source for direct/industrial and indirect portable usages (PUB, 2019), while China has committed to wastewater reuse to release the increasing pressure on water supply (Lyu *et al.*, 2016).

Generally, the secondary effluent from the conventional biological treatment of municipal wastewater is further reclaimed through a number of barriers, including microfiltration (MF)/ultrafiltration (UF), reverse osmosis (RO) and ultraviolet disinfection for NEWater production. However, this process is facing some challenges, e.g. relatively high energy consumption, excessive sludge generation in the conventional biological process etc. Recently, an innovative integrated anaerobic membrane bioreactor (AnMBR) and RO process has been proposed to reclaim wastewater to high-grade product water (Gu *et al.*, 2019; Zhang *et al.*, 2019c). As illustrated in Figure 4.5, an AnMBR was employed as the lead A-stage for energy recovery from municipal wastewater through direct COD capture with minimized sludge generation, whereas RO as the B-stage serves as a barrier to remove nutrients and major ions for production of high-grade reclaimed water.

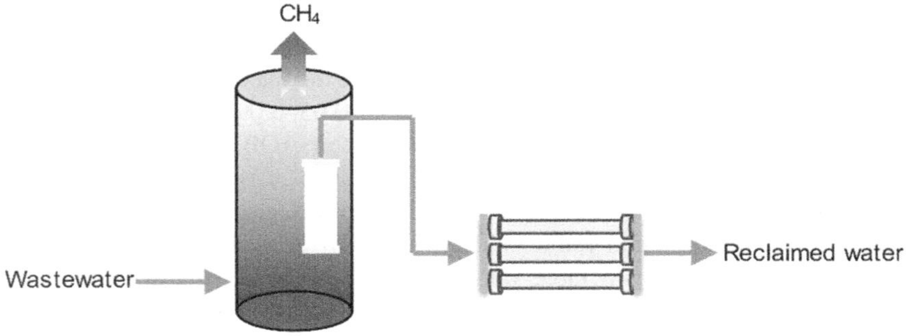

Figure 4.5 Schematic diagram of the integrated AnMBR-RO process. Adapted from Gu *et al.* (2019), with permission.

4.3.1 Performance of integrated AnMBR-RO process

So far, the feasibility of an integrated AnMBR-RO process for wastewater reclamation has been demonstrated in a lab-scale facility. Gu *et al.* (2019) reported a novel integrated AnMBR and RO process for treating synthetic municipal wastewater containing 400 mg/L COD, 45 mg/L NH_4^+-N and 5 mg/L PO_4^{3-}-P. An AnMBR equipped with hollow fibre polyvinylidene fluoride membranes with a pore size of 0.02 μm was operated at an HRT of 6.4 h with a flux at 10 L/(m^2·h), and the AnMBR permeate was further treated by an RO unit operated at a flux of 20 L/(m^2·h). As summarized in Table 4.3, during the operation at steady-state, 95.6% of influent COD could be removed in the AnMBR, of which 76.8% was converted to biomethane with a methane yield of 0.28 L/g COD. More than 95% of the residual organic matter, ammonium and phosphate were rejected by the subsequent RO unit. Meanwhile, the concentrations of the major ions (e.g. calcium, magnesium, potassium, iron) were all below 0.1 mg/L in the RO permeate with a rejection rate of more than 96%. After further polishing of the RO permeate by an ion exchanger (IE), the final product water was capable of meeting or even exceeding the strict effluent discharge standard in China and typical NEWater quality in Singapore in terms of organic matter, ammonia, nitrite, nitrate, phosphate, major ions, etc.

Membrane fouling control is a key issue that determines the performance of the integrated AnMBR-RO process. Biogas sparging, particle sparging and rotating membranes have been proposed as three typical scouring methods for fouling control in an AnMBR. In a study by Wu *et al.* (2017), fluidized granular activated carbon was used to mitigate the issue of membrane fouling of AnMBRs owing to the mechanical scouring and adsorption of soluble organic matters. Biogas sparging, with a sparging rate of 0.15–1.22 m^3/(h·m^2), was also reported for membrane fouling control in AnMBRs (Shin & Bae, 2018). Moreover, with

Table 4.3 Effluent quality in the integrated AnMBR-RO process. Data from Gu *et al.* (2019).

Parameter, mg/L	AnMBR Permeate	RO Permeate	IE Effluent	Quasi Class IV Standards of Surface Water (China)[a]	NEWater (Singapore)[b]
TOC	3.60	0.13	0.13	–	0.50
COD	–	–	–	30	
Ammonium-N	41.90	2.10	<1.00	1.5	1.00
Total nitrogen	–	–	–	10–15	–
Phosphate-P	4.41	0.03	0.03	0.3	–
Calcium	31.20	0.05	<0.01	–	1.00
Iron	0.33	<0.005	<0.005	–	0.04
Sodium	132.70	3.20	3.50–6.60	–	20.00
Potassium	10.18	0.084	<0.01	–	–
Sulphate	31.90	0.50	0.50	–	5.00
Chloride	157.80	4.70	4.70	–	20.00
Conductivity, μS/cm	1127.00	47.00	<39.00	–	100.00

[a]Information from MEEP Beijing (2012); EEB Tianjin (2015).
[b]Information from PUB (2017).

the addition of a coagulant, fouling potential could be substantially reduced through the formation of large flocs in the bulk liquid (Judd & Judd, 2010).

4.3.2 Energy analysis of the integrated AnMBR-RO process

According to a survey by Shin and Bae (2018), the energy consumption in a large-scale AnMBR with a specific gas sparging rate of 0.23 $m^3/(h \cdot m^2)$ was estimated at about 0.21 kWh/m^3, of which membrane scouring might account for about 70% of the total energy consumption. On the other hand, the typical energy consumption by RO had been reported at about 0.57 kWh/m^3 product water (Lee *et al.*, 2013). Therefore, the total energy consumption for producing the high-grade reclaimed water via the AnMBR-RO process was supposed to be 0.85 kWh/m^3 product water at an RO recovery rate of 75%. However, a methane yield of 0.28 L/g COD was achievable in the integrated AnMBR-RO process (Gu *et al.*, 2019). For a municipal wastewater with typical COD concentration of 500 mg/L, a total of 0.14 m^3 of methane could be generated from one cubic meter of wastewater in an AnMBR. Moreover, the COD recovered in the form of

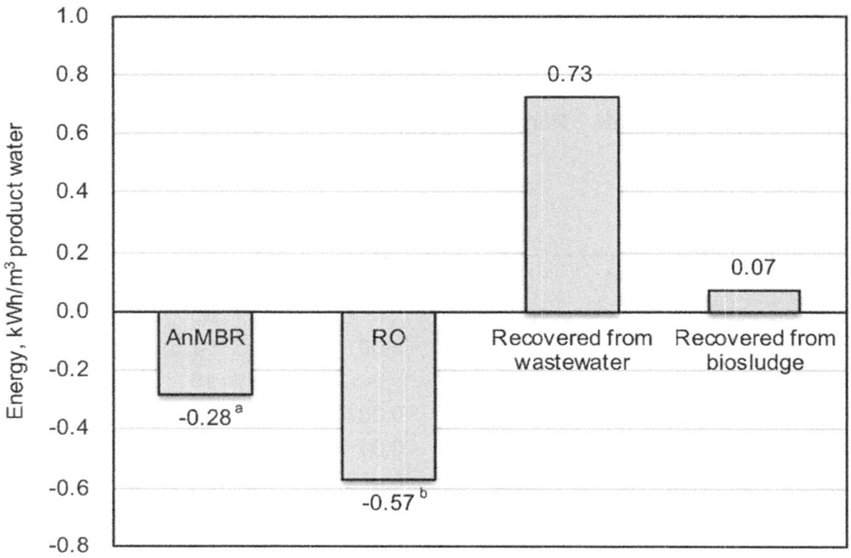

Figure 4.6 Energy consumption and recovery potential in the integrated AnMBR-RO process for wastewater reclamation. [a] Data from Shin and Bae (2018), [b] Data from Lee *et al.* (2013).

biosludge might account for 8% of influent COD in an anaerobic process (Wan *et al.*, 2016). Given the energy potential of 40 MJ/m^3 methane and 13.91 kJ/g methane-COD (Tchobanoglous *et al.*, 2003), the potential recoverable energy could be estimated to be 0.14 m^3 methane $\times$ 40 MJ/m^3 methane + 13.91 kJ/g methane COD $\times$ 500 g COD/m^3 treated wastewater $\times$ 0.08 = 6.16 MJ/m^3 treated wastewater, equivalent to 0.60 kWh/m^3 of electric energy if the conversion efficiency of chemical to electric energy is assumed to be 35%.

As shown in Figure 4.6, the total energy potentially recovered from biomethane and biosludge would reach 0.80 kWh/m^3 product water at an RO recovery rate of 75%, and could roughly offset 94% of the energy consumed for water production in the integrated AnMBR-RO process, indicating a net energy consumption of 0.05 kWh/m^3 product water. This suggests that there is a great possibility of achieving energy self-sufficient or even energy positive municipal wastewater reclamation by adopting innovative process configurations.

4.3.3 Engineering implications

Municipal wastewater should be regarded as a source of fresh water, chemical energy and various resources, rather than viewed as "waste". Indeed, the integrated AnMBR-RO process (Figure 4.5) offers new insights into the possible routes towards sustainable wastewater reclamation.

As discussed above, municipal wastewater could be reclaimed to high grade product water in the AnMBR-RO process, offering an alternative solution for alleviating water shortage. In Singapore, reclaimed ultra-clean high-grade water is used for various industrial or indirect potable uses. In China, the annual production of municipal wastewater already reached 53.2 billion m^3 in 2015 (China MEP, 2017). If the municipal wastewater produced in China was all reclaimed to ultrapure water by adopting the AnMBR-RO process for industrial uses, this would help to significantly save the use of natural surface and ground waters.

Strengthened effluent discharge standards have been implemented in more and more countries due to the increasing concerns about environmental sustainability, which inevitably leads to process upgrading and retrofitting. For instance, in China, many various post-treatment technologies are being adopted to polish the secondary effluent from current biological treatment processes (Figure 4.7).

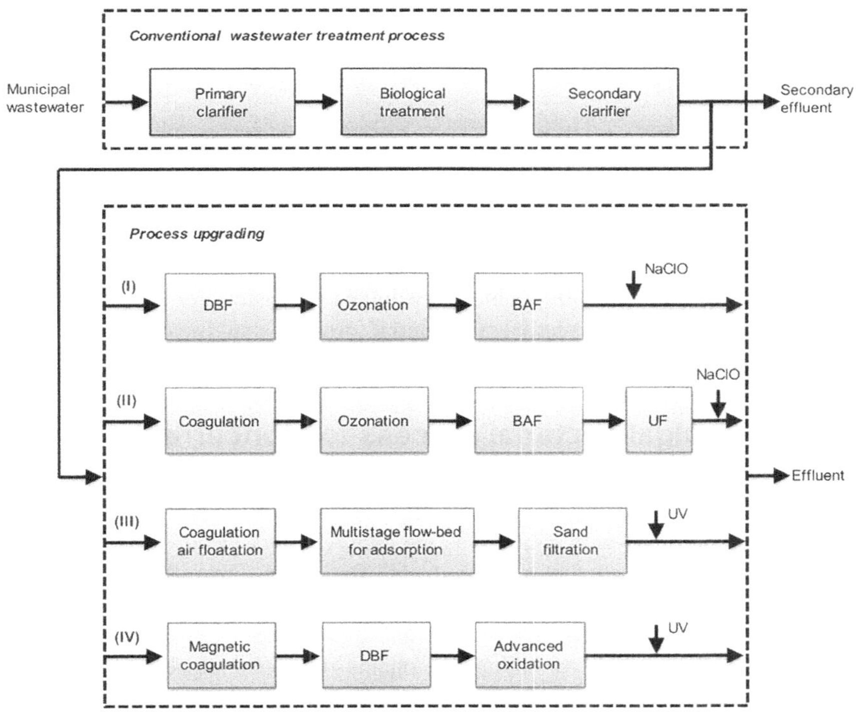

Figure 4.7 Upgrading of the conventional wastewater treatment process to meet the stringent discharge standards of China. DBF: Denitrification biological filter; BAF: Biological aerated filter; UF: Ultrafiltration. Adapted from Zhang *et al.* (2019c), with permission.

However, these efforts in turn make the whole treatment system very complicated and costly with an enlarged footprint (Zhang *et al.*, 2019c). As presented in Table 4.3, the proposed AnMBR-RO process (Figure 4.5) provides an alternative to generate high-quality product water which is capable of meeting strict effluent discharge standards, while with simplified process configuration and reduced footprint. Moreover, primary and secondary clarifiers are no longer required due to the integration of a membrane into the anaerobic process, thus making a reduction of about 52% in the overall footprint achievable in the AnMBR-RO process (Gu *et al.*, 2019), relieving the pressure on limited land space in highly urbanized countries.

Compared with the conventional biological treatment process, waste sludge production in the AnMBR-RO process (Figure 4.5) can be significantly reduced due to the direct conversion of most COD into methane. In fact, the total sludge production yield was reported to be at only 0.06 g VSS/g COD in the AnMBR, indicating a reduction of about 88% in sludge production compared to that in the CAS process with a typical value of 0.3–0.5 g VSS/g COD (Gu *et al.*, 2018). Significantly reduced sludge production helps to largely reduce the cost for sludge post treatment and alleviates the demand on land for waste sludge disposal etc.

4.4 MICROALGAL-BACTERIAL PROCESS

The microalgal-bacterial process has attracted increasing research interest worldwide due to its lower energy demand, cost-effective nutrient removal and potential resource recovery. Life cycle analysis suggests that algae cultivation would be highly economically viable when linked to wastewater treatment (Yang *et al.*, 2011). Thus, the microalgal-bacterial process can be considered as a sustainable platform for municipal wastewater treatment.

4.4.1 Microalgal-bacterial process for concurrent wastewater treatment and energy and resource recovery

Microalgae known as photoautotrophic microbes are able to take up carbon dioxide and inorganic nutrients with sunlight as the energy source, while producing algal biomass and oxygen as the end products. By contrast, in the conventional municipal wastewater treatment process, COD and nutrients (N and P) are converted to carbon dioxide, nitrogen gas and biomass at the cost of high energy consumption. According to the food chains between algae and bacteria, microalgae and bacteria can ideally form a mutually beneficial symbiosis, i.e. microalgae provide oxygen for bacteria in the presence of carbon dioxide and light, and the produced oxygen can then be utilized by bacteria to oxidize organic matters in wastewater, while producing carbon dioxide for microalgal photosynthesis, as illustrated in Figure 4.8. This idea led to the development of

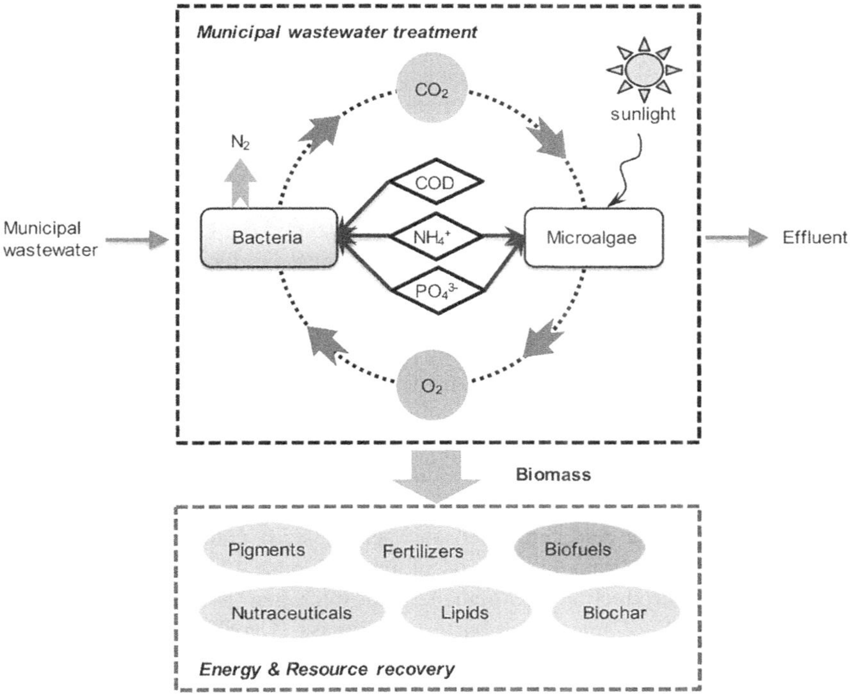

Figure 4.8 Conceptual integrated microalgal-bacterial process for concurrent wastewater treatment and energy & resource recovery. Adapted from Quijano *et al.* (2017), with permission.

an integrated algal-bacterial process for cost-effective municipal wastewater treatment.

Concurrent wastewater treatment and algal biomass production make the simultaneous energy generation and nutrient (N and P) removal possible. In this coupled process, most organic matter is utilized by bacterial respiration for carbon dioxide production, while ammonia and phosphorus are mainly removed via microbial assimilation by microalgae and bacteria (Quijano *et al.*, 2017). In general, the oxygen demand for bacteria to oxidize one unit of COD is equal to the amount of oxygen produced in the chloroplast of microalgae, with carbon dioxide being produced from the oxidation of one unit of COD (Equation 4.4) (Garrett & Grisham, 2008):

$$12H_2O + 6CO_2 + \text{light energy} \rightarrow C_6H_{12}O_6 + 6O_2 + 6H_2O \quad (4.4)$$

As such, no external oxygen is needed for COD removal from wastewater, leading to a significant energy saving associated with aeration in the CAS process for municipal wastewater treatment. In addition, carbon dioxide produced from mineralized COD is recycled and immobilized into the biomass of the microalgae. Obviously, this is beneficial for carbon neutral and environmentally sustainable municipal wastewater treatment.

In addition, the biomass harvested from the integrated microalgal-bacterial process can be used for biodiesel production. For example, it has been predicted that biofuels from microalgae will account for more than 20% of the Italian biofuel market by 2030 (Gambelli *et al.*, 2017). As the third-generation technology for biodiesel production, microalgae-based biodiesel is obviously advantageous over the others, with high oil content and lipid yield. It has been reported that centrate wastewater (from the dewatering of sludge) supplemented with crude glycerol was used for lipid production in a pilot-scale microalgal-bacterial process (Ren *et al.*, 2017). In this system, it was found that 460 mg/(L·d) biomass could be harvested with a lipid content of 27%, together with more than 86% nutrient removal. A higher lipid content of 32.7% was also observed in microalgae biomass cultivated with municipal wastewater (Lam *et al.*, 2017). Moreover, other high-value products, such as pigments, fertilizers, nutraceuticals (e.g. eicosapentaenoic acid and docosahexaenoic acid) could also be obtained from the harvested microalgal-bacterial biomass (Figure 4.8) (Gao *et al.*, 2014). In addition, the production of methane and hydrogen is achievable through anaerobic digestion of microalgal-bacterial biomass (Carrillo-Reyes & Buitrón, 2016). Consequently, it is reasonable to consider that the microalgal-bacterial process could offer an economically viable and environmentally friendly technology for energy-efficient municipal wastewater treatment.

4.4.2 Microalgal-bacterial granules technology

It should be noted that in the suspended microalgal-bacterial process, biomass harvesting is a major challenge due to the poor settleability of the algal biomass resulting from its small cell size, relatively low cell density and the negative charge of the microalgal cell surface. This in turn substantially increases the cost of biomass harvesting (Quijano *et al.*, 2017). It had been estimated that biomass harvesting may account for more than 30% of the total cost of biomass production (Grima *et al.*, 2003). So far, many biomass harvesting methods have been developed, including filtration, centrifugation, chemical flocculation and its hybrid processes, while none of them appears to be economically feasible for large-scale applications (Ma *et al.*, 2018).

Microalgal-bacterial granules, a combination of microalgal floc and bacterial granules, may provide a promising alternative for solid-liquid separation due to their excellent settleability by gravity sedimentation. As illustrated in Figure 4.9, microbial-bacterial granules possess a spatial layered structure with various functional zones for simultaneous and synergetic removal of COD and nutrients in a single unit. Microalgae situating at the out-layer receive sunlight and carbon

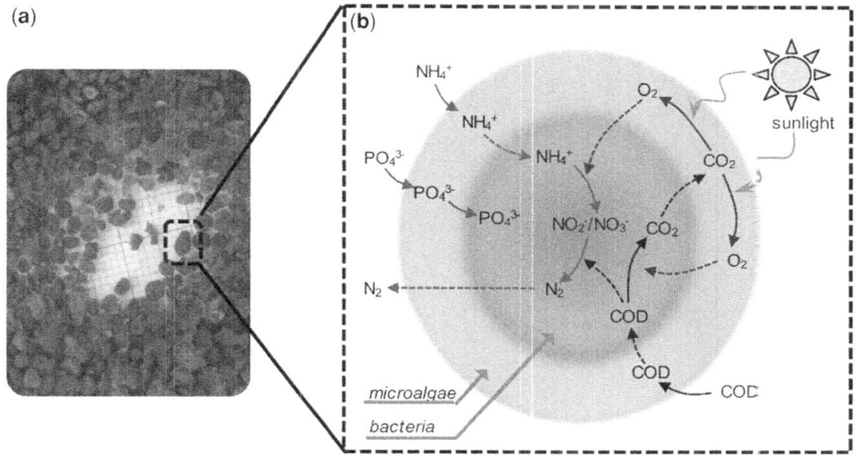

Figure 4.9 Simplified schematic description of microalgal-bacterial granules. (a): Cultivated microalgal-bacterial granules; (b): Illustration of COD and nutrient removal in layer-structured microalgal-bacterial granules.

dioxide from bacteria located inside the granules, while generating oxygen that is readily utilized by bacteria for oxidizing COD to produce new biomass and carbon dioxide. Apart from the assimilation of nutrients to biomass, ammonia can also be removed through nitrification and denitrification in the presence of oxygen and COD in algal-bacterial granules (Karya *et al.*, 2013).

Microalgal-bacterial granules can be obtained under aerated conditions and non-aerated conditions in the presence of light, and are called activated algal granules and photo-granules, respectively (Abouhend *et al.*, 2018; Tiron *et al.*, 2017). During the formation of photo-granules, extracellular polymeric substances secreted by microalgae play a key role, with a higher ratio of polysaccharides to proteins (Kuo-Dahab *et al.*, 2018). Shear force could accelerate granule formation in stirred open systems or aerated reactors (Liu & Tay, 2002). Mobile filamentous cyanobacteria found in a microalgal-bacterial granule has a positive effect on the granule formation through its gliding motility accompanied by a continuous secretion of slime or extracellular polymeric substances, which can wrap heterotrophic bacteria in the closed ecosystem (Milferstedt *et al.*, 2017). Moreover, the excessive growth of filamentous cyanobacteria can form an interwoven mat favouring bio-granulation (Kuo-Dahab *et al.*, 2018). Typically, the size of microalgal-bacterial granules ranges from 0.6 to 3.1 mm in diameter, and the settling velocity is around 14.6–21.6 m/h (Table 4.4). It should be noted that aerobic microalgal-bacterial granules usually have a higher biomass density than those cultivated under non-aerobic conditions.

Table 4.4 Performances and profiles of microalgal-bacterial granules in wastewater treatment.

Wastewater Characteristics, mg/L	Reactor Volume, L	Light Intensity, μmol/(m^2·s)	Aeration Yes/No	Light-Dark Cycle h		HRT h	Performance, %	Lipid Content, mg/g SS	Granule Size, mm	Settling Velocity, m/h	References
				Light Phase	Dark Phase						
COD: 100–400 NH_4^+-N: 8.7–42 PO_4^{3-}-P: 0.8–12	1.5	235	No	15	9	24	COD: 86–98 NH_4^+-N: 53–74 PO_4^{3-}-P: 11–85	N.A.	0.6–1.9	19–22	Tiron *et al.* (2017); Tiron *et al.* (2015)
COD: 90–300 TDN: 23–39 TP: 3.4 ± 0.7	1.5	150	No	2.5	3.5	18–21.6	COD: 77–86 NH_4^+-N: 90–96 TN: 41–57 PO_4^{3-}-P:38–44	N.A.	0.8–2.9	14.6 ± 0.5	Abouhend *et al.* (2018); Milferstedt *et al.* (2017)
COD: 816 NH_4^+-N: 64 TN: 110 PO_4^{3-}-P: 15	50	240	No	N.A.[a]	N.A.[a]	240	COD: 89 TN: 60 PO_4^{3-}-P: 28	N.A.	1.3	18.1 ± 2.1	Arcila and Buitrón (2017)
COD: 600 NH_4^+-N: 100 PO_4^{3-}-P: 10	1.4	Sunlight[b]	Yes	20	4	8	COD: 96 NH_4^+-N: 98 TN: 41–45 TP: 44	N.A.	0.78	N.A.	Huang *et al.* (2015)
COD: 300, NH_4^+-N: 35 PO_4^{3-}-P: 10–11	2	81	Yes	12	12	7–7.5	NH_4^+-N: 93 TN: 50–55 PO_4^{3-}-P: 35	47.9	0.6–1.5	N.A.	Liu *et al.* (2017); Liu *et al.* (2018)
COD: 600 NH_4^+-N: 50 PO_4^{3-}-P: 10	2	135	Yes	12	12	8	COD: 95 NH_4^+-N: >99 TIN: 50–72 TP: 35–38	46.3	3.1	N.A.	Meng *et al.* (2019)

N.A.: not available.
[a]Duration of light and dark phases depends on the practical conditions of outdoor experiments.
[b]Light density depends on sunlight, without control.

High rate algal ponds (HRAPs) have been used for organic matter and nutrients removal from municipal wastewater since the 1970s (Oswald, 1973). Sequential batch HRAPs operated in a sequencing mode are capable of selecting microalgal and bacterial species able to aggregate and settle, while washing out the rest of the microbes in suspension. To date, HRAPs are the most popular system for municipal wastewater treatment. As shown in Table 4.4, microalgal-bacterial granules can remove COD and nutrients from municipal wastewater without aeration, while a better performance at a shorter HRT is achievable when aeration is available. The average removal efficiencies of COD, total nitrogen (TN), NH_4^+-N and total phosphorus (TP) in aerated microalgal-bacterial granules systems can reach 95.9%, 55.4%, 97.3% and 35.8% at an HRT of 8.5 h, respectively, while 87.4%, 54.5%, 78.3% and 41.7% (respectively) are achievable in non-aerated microalgal-bacterial granules systems operated at a longer HRT of 22 h (Table 4.4). In fact, such performance is comparable or even better than that obtained in biological wastewater treatment processes.

Several factors may affect the performance of microalgal-bacterial granules in treating municipal wastewater, such as light intensity, aeration and wastewater characteristics:

- **Light intensity:** light is the energy source of algal growth and it significantly affects the formation of microalgal-bacterial granules. It was observed that no microalgal-bacterial granules were formed in a HRAP when the solar irradiance level reached 545 μmol/(m^2·s) (Arcila & Buitrón, 2017). By contrast, a slightly lower irradiance could accelerate the formation of granules through enhanced production of extracellular polymeric substances. Generally, the favourable photosynthetic photon flux density of illumination for microalgal-bacterial granules formation is in the range of 80–330 μmol/(m^2·s) (Arcila & Buitrón, 2017; Quijano *et al.*, 2017). In the integrated algal-bacterial process, the light intensity determines the production of oxygen by microalgae which is subsequently utilized by bacteria, thus it ultimately may affect the performance of wastewater treatment. For example, it has been reported that nutrient removal was improved by increasing light intensity from 45 to 225 μmol/(m^2·s) (Meng *et al.*, 2019). However, intense light may also trigger microbial successions in both microalgae and bacteria related to wastewater treatment. It was observed that the abundance of *Cyanobacteria* and *Chlorophyta* tended to increase with light intensity, and a similar trend was also observed for some functional bacterial species, e.g. *Saprospiraceae* for protein hydrolysis, *Comamonadaceae* for phosphorus removal and *Nitrosomonadaceae* for nitrification (Meng *et al.*, 2019). It should be noted that a light intensity higher than180 μmol/(m^2·s) might be unfavourable for the growth of some nitrite-oxidizing bacteria (e.g. *Nitrospiraceae*), leading to the accumulation of nitrite (e.g. up to 30–50% of influent ammonia) in treated effluent (Huang *et al.*, 2015; Liu *et al.*, 2018). Thus, it is a reasonable consideration that microalgal-bacterial granules technology could be integrated with the anammox process due to the considerable nitrite accumulation. Finally, it should be noted that increased light intensity definitely leads to higher energy consumption, and a balance between the performance and costs should be carried out.

- **Aeration:** aeration is helpful for fast granulation and for better performance of wastewater treatment. As shown in Table 4.4, the average removal efficiencies of COD and NH_4^+-N in aerated microalgal-bacterial granules processes were increased by 8.5% and 19%, respectively, compared with those in the non-aerated processes due to the contribution of aerobic heterotrophic bacteria. However, the non-aerated processes have the advantages of low energy consumption and reduced loss of carbon dioxide. A complete life cycle analysis of these two algal-bacterial processes should be performed in a holistic manner in terms of economic viability, engineering feasibility and environmental sustainability when applied for municipal wastewater treatment.
- **Wastewater characteristics:** apart from organic matter and nutrients, municipal wastewater usually contains various ions, such as divalent cations and inorganic carbon, which may affect the formation of microalgal-bacterial granules and their performance. For example, the addition of divalent cations (e.g. Ca^{2+} and Mg^{2+}) can benefit the structural stability of microalgal-bacterial granules by mediating multiple cross-linkages among polysaccharides, sugars and protein chains in extracellular polymeric substances (Xiao & Zheng, 2016). Concentrations of Ca^{2+} and Mg^{2+} of 80 and 190 mg/L, respectively, have been reported to promote the aggregation of microalgal and bacterial cells (Powell & Hill, 2013). In general, concentrations of Ca^{2+} and Mg^{2+} in municipal wastewater are at the level of around 55 and 15 mg/L, respectively (Schönborn *et al.*, 2001), which may also depend on the hardness of local source water. Moreover, it has been observed that microbial-bacterial granules grown on wastewater with a low inorganic/organic carbon ratio exhibited high settleability and nitrogen removal efficiency (Van Den Hende *et al.*, 2011). It appears that adding inorganic carbon (e.g. bicarbonate) to wastewater to enhance the performance of the microalgal-bacterial process has debatable advantages. Thus, work is needed to obtain more relevant information for the proper design, operation and optimization of microalgal-bacterial processes.

The microalgal-bacterial process is built on the basis of the synergetic reactions between microalgae and bacteria, i.e. microalgae produce oxygen which can be used by aerobic bacteria to oxidize organic matter, while the carbon dioxide produced is taken up by the microalgae through photosynthetic reactions. In such a coupled process, most of the ammonia and phosphorus are simultaneously removed from wastewater via microbial assimilations. In addition, microalgal-bacterial biomass can be harvested for further processing with the aim of producing value-added products, such as biodiesels, pigments and fertilizer. The harvested biomass can also be subjected to anaerobic digestion for biogas production. For example, average methane production with dispersed biomass has been reported at about 227 $\pm$ 57 ml CH_4/g VSS, while it was 186 $\pm$ 94 ml CH_4/g VSS for granular biomass (Quijano *et al.*, 2017), which could be respectively increased by 60% and 72% after proper pretreatment. These are indeed comparable with or even better than that from the anaerobic digestion of activated sludge (e.g. 240–367 ml CH_4/g VSS) (Yin *et al.*, 2016). In WWTPs, aeration associated energy consumption accounts for about 50% of in-plant energy consumption, while the recoverable electric energy from anaerobic digestion of waste sludge could only offset about 50–60% of the total energy consumption. In contrast, aeration associated energy can be saved in the microbial-bacterial process, while energy can be recovered from the anaerobic digestion of microalgal-bacterial biomass. In this regard, the microbial-bacterial process may offer a promising option for energy self-sufficient wastewater treatment.

In China, the average influent concentrations of COD, NH_4^+-N and PO_4^{3-}-P are about 220, 30 and 3.7 mg/L, respectively, based on a survey of 127 WWTPs (Guo *et al.*, 2018). As discussed above, if 95.9%, 97.3% and 35.8% of COD, NH_4^+-N and PO_4^{3-}-P removal (respectively) could be achieved in the microalgal-bacterial granules process, the effluent should have a quality of 8.8 mg/L COD, 0.81 NH_4^+-N and 2.4 mg/L PO_4^{3-}-P, respectively. These suggest that effluent COD and NH_4^+-N easily meet the discharge standard of Grade I-A in China, while the effluent PO_4^{3-}-P concentration may not. In fact, phosphorus is mainly removed through biomass assimilation in the algal-bacterial process. According to the elemental composition of microalgal-bacterial granules, e.g. $C_{106}H_{263}O_{110}N_{10}P_1$ (Abinandan *et al.*, 2018), phosphorus in wastewater usually exceeds the required amount for the growth of microalgal-bacterial granules. As such, a polishing step of effluent from the algal-bacterial process is still needed, e.g. chemical precipitation of phosphate. Moving forward, microalgal-bacterial technology could also be integrated with an RO unit to achieve an innovative A-B process for NEWater production, where microalgal-bacterial technology is employed as the lead A-stage to obtain the low-strength effluent, whereas RO is used as the B-stage, serving as a barrier to remove nutrients and major ions for the production of high-grade reclaimed water.

4.5 CONCLUSIONS

This chapter presents several emerging biological processes for municipal wastewater treatment with the aim of improving process energy efficiency. It appears that mainstream deammonification in the configurations of PN-anammox and PDN-anammox likely offers a great opportunity for energy-neutral municipal wastewater treatment. The recently developed integrated AnMBR-RO process has also exhibited the advantages of high energy efficiency, significantly reduced sludge production, high-grade product water, small footprint, etc., and the microalgal-bacterial process provides another alternative platform for municipal wastewater treatment with lower energy demand, enhanced nutrient removal and potential resource recovery. Though these emerging biological technologies are still at their developmental stages, they show possible directions for next-generation biological processes in municipal WWTPs, which should be designed as holistic water, energy and resource factories.

REFERENCES

Abinandan S., Subashchandrabose S. R., Venkateswarlu K. and Megharaj M. (2018). Nutrient removal and biomass production: advances in microalgal biotechnology for wastewater treatment. *Critical Reviews in Biotechnology*, **38**(8), 1244–1260.

Abouhend A. S., McNair A., Kuo-Dahab W. C., Watt C., Butler C. S., Milferstedt K., Hamelin J., Seo J., Gikonyo G. J. and El-Moselhy K. M. (2018). The oxygenic photogranule process for aeration-free wastewater treatment. *Environmental Science & Technology*, **52**(6), 3503–3511.

Ahn J. H., Yu R. and Chandran K. (2008). Distinctive microbial ecology and biokinetics of autotrophic ammonia and nitrite oxidation in a partial nitrification bioreactor. *Biotechnology and Bioengineering*, **100**(6), 1078–1087.

Anthonisen A., Loehr R., Prakasam T. and Srinath E. (1976). Inhibition of nitrification by ammonia and nitrous acid. *Journal (Water Pollution Control Federation)*, **48**(5), 835–852.

Arcila J. S. and Buitrón G. (2017). Influence of solar irradiance levels on the formation of microalgae-bacteria aggregates for municipal wastewater treatment. *Algal Research*, **27**, 190–197.

Boccaletti G., Grobbel M. and Stuchtey M. R. (2010). The business opportunity in water conservation. *McKinsey Quarterly*, **1**, 67–75.

Bournazou M. C., Hooshiar K., Arellano-Garcia H., Wozny G. and Lyberatos G. (2013). Model based optimization of the intermittent aeration profile for SBRs under partial nitrification. *Water Research*, **47**(10), 3399–3410.

Cao Y. S., Kwok B., Yong W., Chua S., Wah Y. and Yahya A. (2013). The main stream autotrophic nitrogen removal in the largest full scale activated sludge process in Singapore: process analysis. *Proceedings of WEF/IWA Nutrient Removal and Recovery 2013: Trends in Resource Recovery and Use*, July 28–31, Vancouver.

Cao S. B., Du R., Li B. K., Wang S. Y., Ren N. Q. and Peng Y. Z. (2017a). Nitrite production from partial-denitrification process fed with low carbon/nitrogen (C/N) domestic wastewater: performance, kinetics and microbial community. *Chemical Engineering Journal*, **326**, 1186–1196.

Cao Y. S., Kwok B. H., van Loosdrecht M. C. M., Daigger G. T., Png H. Y., Long W. Y., Chye C. S. and Ghani Y. (2017b). The occurrence of enhanced biological phosphorus removal in a 200,000 m^3/day partial nitration and Anammox activated sludge process at the Changi water reclamation plant, Singapore. *Water Science and Technology*, **75**(3), 741–751.

Cao Y. S., van Loosdrecht M. C. M. and Daigger G. T. (2017c). Mainstream partial nitritation-anammox in municipal wastewater treatment: status, bottlenecks, and further studies. *Applied Microbiology and Biotechnology*, **101**(4), 1365–1383.

Carrillo-Reyes J. and Buitrón G. (2016). Biohydrogen and methane production via a two-step process using an acid pretreated native microalgae consortium. *Bioresource Technology*, **221**, 324–330.

Chen J. and Strous M. (2013). Denitrification and aerobic respiration, hybrid electron transport chains and co-evolution. *Biochimica et Biophysica Acta (BBA)-Bioenergetics*, **1827**(2), 136–144.

China Ministry of Environmental Protection (China MEP) (2017). National Report on Environmental Quality of 2015. China MEP, China.

Daigger G. T. (2014). Oxygen and carbon requirements for biological nitrogen removal processes accomplishing nitrification, nitritation, and anammox. *Water Environment Research*, **86**(3), 204–209.

Daims H., Nielsen J. L., Nielsen P. H., Schleifer K.-H. and Wagner M. (2001). In situ characterization ofnitrospira-like nitrite-oxidizing bacteria active in wastewater treatment plants. *Applied and Environmental Microbiology*, **67**(11), 5273–5284.

Du R., Cao S. B., Li B. K., Niu M., Wang S. Y. and Peng Y. Z. (2017). Performance and microbial community analysis of a novel DEAMOX based on partial-denitrification and anammox treating ammonia and nitrate wastewaters. *Water Research*, **108**, 46–56.

Du R., Peng Y., Cao S., Li B., Wang S. and Niu M. (2016a). Mechanisms and microbial structure of partial denitrification with high nitrite accumulation. *Applied Microbiology and Biotechnology*, **100**(4), 2011–2021.

Du R., Peng Y. Z., Cao S. B., Wang S. Y. and Niu M. (2016b). Characteristic of nitrous oxide production in partial denitrification process with high nitrite accumulation. *Bioresource Technology*, **203**, 341–347.

Du R., Cao S., Li B., Zhang H., Wang S. and Peng Y. (2019a). Synergy of partial-denitrification and anammox in continuously fed upflow sludge blanket reactor for simultaneous nitrate and ammonia removal at room temperature. *Bioresource Technology*, **274**, 386–394.

Du R., Cao S., Peng Y., Zhang H. and Wang S. (2019b). Combined Partial Denitrification (PD)-Anammox: a method for high nitrate wastewater treatment. *Environment International*, **126**, 707–716.

Ecological and Environmental Bureau, Tianjin (EEB, Tianjin). (2015). Discharge Standard of Water Pollutants for Municipal Wastewater Treatment Plant. (DB12/599–2015). Available at: http://sthj.tj.gov.cn/root16/mechanism/standard_monitoring_of_the_Department_of_science_and_technology/201509/t20150928_17652.html (accessed 12 October 2019).

Fux C., Velten S., Carozzi V., Solley D. and Keller J. (2006). Efficient and stable nitritation and denitritation of ammonium-rich sludge dewatering liquor using an SBR with continuous loading. *Water Research*, **40**(14), 2765–2775.

Gambelli D., Alberti F., Solfanelli F., Vairo D. and Zanoli R. (2017). Third generation algae biofuels in Italy by 2030: a scenario analysis using Bayesian networks. *Energy Policy*, **103**, 165–178.

Gao H., Scherson Y. D. and Wells G. F. (2014). Towards energy neutral wastewater treatment: methodology and state of the art. *Environmental Science: Processes & Impacts*, **16**(6), 1223–1246.

Garrett R. and Grisham C. (2008). *Biochemistry* (4th ed.). Brooks/Cole, Cengage Learning, Boston, USA.

Glass C. and Silverstein J. (1998). Denitrification kinetics of high nitrate concentration water: pH effect on inhibition and nitrite accumulation. *Water Research*, **32**(3), 831–839.

Gong L., Huo M., Yang Q., Li J., Ma B., Zhu R., Wang S. and Peng Y. (2013). Performance of heterotrophic partial denitrification under feast-famine condition of electron donor: a case study using acetate as external carbon source. *Bioresource Technology*, **133**, 263–269.

Grima E. M., Belarbi E.-H., Fernández F. A., Medina A. R. and Chisti Y. (2003). Recovery of microalgal biomass and metabolites: process options and economics. *Biotechnology Advances*, **20**(7–8), 491–515.

Gu J., Yang Q. and Liu Y. (2018). Mainstream anammox in a novel A-2B process for energy-efficient municipal wastewater treatment with minimized sludge production. *Water Research*, **138**, 1–6.

Gu J., Liu H., Wang S., Zhang M. and Liu Y. (2019). An innovative anaerobic MBR-reverse osmosis-ion exchange process for energy-efficient reclamation of municipal wastewater to NEWater-like product water. *Journal of Cleaner Production*, **230**, 1287–1293.

Guo J.-H., Peng Y.-Z., Peng C.-Y., Wang S.-Y., Chen Y., Huang H.-J. and Sun Z.-R. (2010). Energy saving achieved by limited filamentous bulking sludge under low dissolved oxygen. *Bioresource Technology*, **101**(4), 1120–1126.

Guo H., Li X., Ren Q., Liu S., Wu Y. and Jia L. (2018). Analysis on characteristics of influent water quality of typical municipal sewage treatment plants in China. *Water & Wastewater (in Chinese)*, **54**(6), 12–15.

Gustavsson D., Persson F. and la Cour Jansen J. (2014). Manammox–mainstream anammox at Sjölunda WWTP. Proceedings from the IWA World Water Congress and Exhibition, September 21–26, Lisbon, Portugal.

Han M., De Clippeleir H., Al-Omari A., Wett B., Vlaeminck S., Bott C. and Murthy S. (2016). Impact of carbon to nitrogen ratio and aeration regime on mainstream deammonification. *Water Science and Technology*, **74**(2), 375–384.

Hellinga C., Schellen A., Mulder J. W., van Loosdrecht M. V. and Heijnen J. (1998). The SHARON process: an innovative method for nitrogen removal from ammonium-rich waste water. *Water Science and Technology*, **37**(9), 135–142.

Henze M., van Loosdrecht M. C., Ekama G. A. and Brdjanovic D. (2008). *Biological Wastewater Rreatment*. IWA Publishing, London, UK.

Huang W., Li B., Zhang C., Zhang Z., Lei Z., Lu B. and Zhou B. (2015). Effect of algae growth on aerobic granulation and nutrients removal from synthetic wastewater by using sequencing batch reactors. *Bioresource Technology*, **179**, 187–192.

Jenni S., Vlaeminck S. E., Morgenroth E. and Udert K. M. (2014). Successful application of nitritation/anammox to wastewater with elevated organic carbon to ammonia ratios. *Water Research*, **49**, 316–326.

Jetten M. S., Horn S. J. and van Loosdrecht M. C. (1997). Towards a more sustainable municipal wastewater treatment system. *Water Science and Technology*, **35**(9), 171–180.

Ji J., Peng Y., Wang B., Mai W., Li X., Zhang Q. and Wang S. (2018). Effects of salinity build-up on the performance and microbial community of partial-denitrification granular sludge with high nitrite accumulation. *Chemosphere*, **209**, 53–60.

Jin L. Y., Zhang G. M. and Tian H. F. (2014). Current state of sewage treatment in China. *Water Research*, **66**, 85–98.

Judd S. and Judd C. I. (2010). The MBR Book: Principles and Applications of Membrane Bioreactors for Water and Wastewater Treatment. Elsevier, Oxford, UK.

Kalyuzhnyi S., Gladchenko M., Mulder A. and Versprille B. (2006). DEAMOX – New biological nitrogen removal process based on anaerobic ammonia oxidation coupled to sulphide-driven conversion of nitrate into nitrite. *Water Research*, **40**(19), 3637–3645.

Kartal B., Kuenen J. V. and Van Loosdrecht M. (2010). Sewage treatment with anammox. *Science*, **328**(5979), 702–703.

Karya N., Van der Steen N. and Lens P. (2013). Photo-oxygenation to support nitrification in an algal–bacterial consortium treating artificial wastewater. *Bioresource Technology*, **134**, 244–250.

Kulikowska D. and Bernat K. (2013). Nitritation–denitritation in landfill leachate with glycerine as a carbon source. *Bioresource Technology*, **142**, 297–303.

Kumar M. and Lin J.-G. (2010). Co-existence of anammox and denitrification for simultaneous nitrogen and carbon removal—strategies and issues. *Journal of Hazardous Materials*, **178**(1–3), 1–9.

Kuo-Dahab W. C., Stauch-White K., Butler C. S., Gikonyo G. J., Carbajal-González B., Ivanova A., Dolan S. and Park C. (2018). Investigation of the Fate and Dynamics of Extracellular Polymeric Substances (EPS) during Sludge-Based Photogranulation under Hydrostatic Conditions. *Environmental Science & Technology*, **52**(18), 10462–10471.

Lackner S., Gilbert E. M., Vlaeminck S. E., Joss A., Horn H. and van Loosdrecht M. C. (2014). Full-scale partial nitritation/anammox experiences–an application survey. *Water Research*, **55**, 292–303.

Lam M. K., Yusoff M. I., Uemura Y., Lim J. W., Khoo C. G., Lee K. T. and Ong H. C. (2017). Cultivation of Chlorella vulgaris using nutrients source from domestic wastewater for biodiesel production: growth condition and kinetic studies. *Renewable Energy*, **103**, 197–207.

Lee Y., Tao G., Lay W., Lin L., Kwok Y. and Koh K. (2013). *Towards Energy Self-Sufficient Water Reclamation Plants*. PUB, Singapore.

Li J., Peng Y., Zhang L., Liu J., Wang X., Gao R., Pang L. and Zhou Y. (2019). Quantify the contribution of anammox for enhanced nitrogen removal through metagenomic analysis and mass balance in an anoxic moving bed biofilm reactor. *Water Research*, **160**, 178–187.

Lin X., Wang Y., Ma X., Yan Y., Wu M., Bond P. L. and Guo J. (2018). Evidence of differential adaptation to decreased temperature by anammox bacteria. *Environmental Microbiology*, **20**(10), 3514–3528.

Liu Y. and Tay J.-H. (2002). The essential role of hydrodynamic shear force in the formation of biofilm and granular sludge. *Water Research*, **36**(7), 1653–1665.

Liu B., Mao Y., Bergaust L., Bakken L. R. and Frostegård Å. (2013a). Strains in the genus Thauera exhibit remarkably different denitrification regulatory phenotypes. *Environtal Microbiology*, **15**(10), 2816–2828.

Liu L., Fan H., Liu Y., Liu C. and Huang X. (2017). Development of algae-bacteria granular consortia in photo-sequencing batch reactor. *Bioresource Technology*, **232**, 64–71.

Liu L., Zeng Z., Bee M., Gibson V., Wei L., Huang X. and Liu C. (2018). Characteristics and performance of aerobic algae-bacteria granular consortia in a photo-sequencing batch reactor. *Journal of Hazardous Materials*, **349**, 135–142.

Liu Y., Ngo H. H., Guo W., Peng L., Wang D. and Ni B. (2019). The roles of free ammonia (FA) in biological wastewater treatment processes: a review. *Environment International*, **123**, 10–19.

Lotti T., Kleerebezem R., Hu Z., Kartal B., de Kreuk M. K., Kip C. V. T., Kruit J., Hendrickx T. L. G. and van Loosdrecht M. C. M. (2015a). Pilot-scale evaluation of anammox-based mainstream nitrogen removal from municipal wastewater. *Environmental Technology*, **36**(9), 1167–1177.

Lotti T., Kleerebezem R. and Van Loosdrecht M. (2015b). Effect of temperature change on anammox activity. *Biotechnology and Bioengineering*, **112**(1), 98–103.

Lyu S., Chen W., Zhang W., Fan Y. and Jiao W. (2016). Wastewater reclamation and reuse in China: opportunities and challenges. *Journal of Environmental Sciences*, **39**, 86–96.

Ma B., Qian W. T., Yuan C. S., Yuan Z. G. and Peng Y. Z. (2017). Achieving Mainstream Nitrogen Removal through Coupling Anammox with Denitratation. *Environmental Science & Technology*, **51**(15), 8405–8413.

Ma Y., Gao Z., Wang Q. and Liu Y. (2018). Biodiesels from microbial oils: opportunity and challenges. *Bioresource Technology*, **263**, 631–641.

Massara T. M., Malamis S., Guisasola A., Baeza J. A., Noutsopoulos C. and Katsou E. (2017). A review on nitrous oxide (N_2O) emissions during biological nutrient removal from municipal wastewater and sludge reject water. *Science of The Total Environment*, **596–597**, 106–123.

Meng F., Xi L., Liu D., Huang W., Lei Z., Zhang Z. and Huang W. (2019). Effects of light intensity on oxygen distribution, lipid production and biological community of algal-bacterial granules in photo-sequencing batch reactors. *Bioresource Technology*, **272**, 473–481.

Metcalf E. and Eddy M. (2014). Wastewater Engineering: Treatment and Resource Recovery. Mic Graw-Hill, USA, 1530–1533.

Milferstedt K., Kuo-Dahab W. C., Butler C. S., Hamelin J., Abouhend A. S., Stauch-White K., McNair A., Watt C., Carbajal-González B. I. and Dolan S. (2017). The importance of filamentous cyanobacteria in the development of oxygenic photogranules. *Scientific Reports*, **7**(1), 17944.

Municipal Ecological Environmental Bureau, Beijing (MEEP, Beijing). (2012). Discharge Standard of Water Pollutants for Municipal Wastewater Treatment Plant (DB11/890–2012). Available at: http://sthjj.beijing.gov.cn/bjhrb/xxgk/fgwj/hjbhbz13/dfhbbz92/9a5cb7c1-2.html (accessed 12 October 2019).

Oswald W. (1973). Productivity of algae in sewage disposal. *Solar Energy*, **15**(1), 107–117.

Park S., Bae W. and Rittmann B. (2010). Operational Boundaries for Nitrite Accumulation in Nitrification Based on Minimum/Maximum Substrate Concentrations That Include Effects of Oxygen Limitation, pH, and Free Ammonia and Free Nitrous Acid Inhibition. *Environmental Science & Technology*, **44**(1), 335–342.

Picioreanu C., Van Loosdrecht M. and Heijnen J. (1997). Modelling the effect of oxygen concentration on nitrite accumulation in a biofilm airlift suspension reactor. *Water Science and Technology*, **36**(1), 147–156.

Powell R. J. and Hill R. T. (2013). Rapid aggregation of biofuel-producing algae by the bacterium Bacillus sp. strain RP1137. *Applied and Environmental Microbiology*, **79** (19), 6093–6101.

PUB (2017). NEWater Quality (Typical value). PUB: Singapore's National Water Agency. Available at: https://www.pub.gov.sg/Documents/PUB_NEWater_Quality.pdf. (Accessed: 22nd January 2019).

PUB (2019). NEWater Usage. PUB: Singapore's National Water Agency. Available at: https://www.pub.gov.sg/watersupply/fournationaltaps/newater (Accessed: 22nd January 2019).

Quijano G., Arcila J. S. and Buitrón G. (2017). Microalgal-bacterial aggregates: applications and perspectives for wastewater treatment. *Biotechnology Advances*, **35** (6), 772–781.

Regmi P., Miller M. W., Holgate B., Bunce R., Park H., Chandran K., Wett B., Murthy S. and Bott C. B. (2014). Control of aeration, aerobic SRT and COD input for mainstream nitritation/denitritation. *Water Research*, **57**, 162–171.

Regmi P., Holgate B., Fredericks D., Miller M. W., Wett B., Murthy S. and Bott C. B. (2015). Optimization of a mainstream nitritation-denitritation process and anammox polishing. *Water Science and Technology*, **72**(4), 632–642.

Reijnders L. (2014). Phosphorus resources, their depletion and conservation, a review. *Resources, Conservation and Recycling*, **93**, 32–49.

Ren H., Tuo J., Addy M. M., Zhang R., Lu Q., Anderson E., Chen P. and Ruan R. (2017). Cultivation of Chlorella vulgaris in a pilot-scale photobioreactor using real centrate wastewater with waste glycerol for improving microalgae biomass production and wastewater nutrients removal. *Bioresource Technology*, **245**, 1130–1138.

Rittmann B. E. and McCarty P. L. (2001). Environmental Biotechnology: Principles and Applications. McGraw-Hill, Boston, c2001.

Roco C. A., Bergaust L. L., Bakken L. R., Yavitt J. B. and Shapleigh J. P. (2017). Modularity of nitrogen-oxide reducing soil bacteria: linking phenotype to genotype. *Environmental Microbiology*, **19**(6), 2507–2519.

Ruiz G., Jeison D., Rubilar O., Ciudad G. and Chamy R. (2006). Nitrification–denitrification via nitrite accumulation for nitrogen removal from wastewaters. *Bioresource Technology*, **97**(2), 330–335.

Schönborn C., Bauer H.-D. and Röske I. (2001). Stability of enhanced biological phosphorus removal and composition of polyphosphate granules. *Water Research*, **35**(13), 3190–3196.

Shapleigh J. P. (2013). Denitrifying prokaryotes. In: The Prokaryotes, E. Rosenberg, E. F. DeLong, S. Lory, E. Stackebrandt and F. Thompson (eds), Springer, Berlin, Heidelberg, pp. 405–425.

Shin C. and Bae J. (2018). Current status of the pilot-scale anaerobic membrane bioreactor treatments of domestic wastewaters: a critical review. *Bioresource Technology*, **247**, 1038–1046.

Si Z., Peng Y. Z., Yang A. M., Zhang S. J., Li B. K., Wang B. and Wang S. Y. (2018). Rapid nitrite production via partial denitrification: pilot-scale operation and microbial community analysis. *Environmental Science-Water Research & Technology*, **4**(1), 80–86.

Strous M., Heijnen J. J., Kuenen J. G. and Jetten M. S. M. (1998). The sequencing batch reactor as a powerful tool for the study of slowly growing anaerobic ammonium-oxidizing microorganisms. *Applied Microbiology and Biotechnology*, **50**(5), 589–596.

Strous M., Kuenen J. and Jetten M. (1999). Key physiology of anaerobic ammonium oxidation. *Applied and Environmental Microbiology*, **65**(7), 3248–3250.

Tchobanoglous G., Burton F. L. and Stensel H. D. (2003). Wastewater Engineering Treatment and Reuse. McGraw-Hill Higher Education, Boston, US.

Tiron O., Bumbac C., Patroescu I., Badescu V. and Postolache C. (2015). Granular activated algae for wastewater treatment. *Water Science and Technology*, **71**(6), 832–839.

Tiron O., Bumbac C., Manea E., Stefanescu M. and Lazar M. N. (2017). Overcoming microalgae harvesting barrier by activated algae granules. *Scientific Reports*, **7**(1), 4646.

Trojanowicz K., Plaza E. and Trela J. (2016). Pilot scale studies on nitritation-anammox process for mainstream wastewater at low temperature. *Water Science and Technology*, **73**(4), 761–768.

Van Den Hende S., Vervaeren H., Saveyn H., Maes G. and Boon N. (2011). Microalgal bacterial floc properties are improved by a balanced inorganic/organic carbon ratio. *Biotechnology and Bioengineering*, **108**(3), 549–558.

Vilpanen M. (2017). Mainstream deammonification–Pre-design for piloting at Viikinmäki wastewater treatment plant. Master thesis, Asltl University, Finland.

Wan J., Gu J., Zhao Q. and Liu Y. (2016). COD capture: a feasible option towards energy self-sufficient domestic wastewater treatment. *Scientific Reports*, **6**, 25054.

Wang H., Xu G., Qiu Z., Zhou Y. and Liu Y. (2019). NOB suppression in pilot-scale mainstream nitritation-denitritation system coupled with MBR for municipal wastewater treatment. *Chemosphere*, **216**, 633–639.

Wett B., Omari A., Podmirseg S., Han M., Akintayo O., Gómez Brandón M., Murthy S., Bott C., Hell M. and Takács I. (2013). Going for mainstream deammonification from bench to full scale for maximized resource efficiency. *Water Science and Technology*, **68**(2), 283–289.

Wu B., Li Y., Lim W., Lee S. L., Guo Q., Fane A. G. and Liu Y. (2017). Single-stage versus two-stage anaerobic fluidized bed bioreactors in treating municipal wastewater: performance, foulant characteristics, and microbial community. *Chemosphere*, **171**, 158–167.

Xiao R. and Zheng Y. (2016). Overview of microalgal extracellular polymeric substances (EPS) and their applications. *Biotechnology Advances*, **34**(7), 1225–1244.

Xu G., Xu X., Yang F., Liu S. and Gao Y. (2012). Partial nitrification adjusted by hydroxylamine in aerobic granules under high DO and ambient temperature and subsequent Anammox for low C/N wastewater treatment. *Chemical Engineering Journal*, **213**, 338–345.

Xu G. J., Zhou Y., Yang Q., Lee Z. M. P., Gu J., Lay W., Cao Y. S. and Liu Y. (2015). The challenges of mainstream deammonification process for municipal used water treatment. *Applied Microbiology and Biotechnology*, **99**(6), 2485–2490.

Yang J., Xu M., Zhang X., Hu Q., Sommerfeld M. and Chen Y. (2011). Life-cycle analysis on biodiesel production from microalgae: water footprint and nutrients balance. *Bioresource Technology*, **102**(1), 159–165.

Yin Y., Liu Y.-J., Meng S.-J., Kiran E. U. and Liu Y. (2016). Enzymatic pretreatment of activated sludge, food waste and their mixture for enhanced bioenergy recovery and waste volume reduction via anaerobic digestion. *Applied Energy*, **179**, 1131–1137.

Zhang M., Zheng P., Li W., Wang R., Ding S. and Abbas G. (2015). Performance of nitrate-dependent anaerobic ferrous oxidizing (NAFO) process: a novel prospective technology for autotrophic denitrification. *Bioresource Technology*, **179**, 543–548.

Zhang M., Gu J. and Liu Y. (2019a). Engineering feasibility, economic viability and environmental sustainability of energy recovery from nitrous oxide in biological wastewater treatment plant. *Bioresource Technology*, **282**, 514–519.

Zhang M., Wang S., Ji B. and Liu Y. (2019b). Towards mainstream deammonification of municipal wastewater: partial nitrification-anammox versus partial denitrification-anammox. *Science of The Total Environment*, **692**, 393–401.

Zhang X., Zhang M., Liu H., Gu J. and Liu Y. (2019c). Environmental sustainability: a pressing challenge to biological sewage treatment processes. *Current Opinion in Environmental Science & Health*, **12**, 1–5.

Chapter 5

Perspectives and actions moving forward

Jun Gu

5.1 INTRODUCTION

The A-B processes have gained great attention for energy-efficient municipal wastewater reclamation. Chapters 3 and 4 clearly demonstrate the feasibility, performance and energy efficiency of A-B processes with various configurations. Analysis shows that A-B processes are capable of enhancing energy recovery through chemical oxygen demand (COD) capture at the A-stage. Compared with the conventional activated sludge (CAS) process, much more influent COD could be harvested as methane, i.e. 58% of total COD could be converted to methane gas via direct anaerobic treatment (Gu *et al.*, 2018). Sludge which is harvested by chemical enhanced primary treatment (CEPT) and the high-rate activated sludge (HRAS) process generally showed higher biodegradability than did biosludge generated by CAS-based processes, leading to a higher methane recovery (Rahman *et al.*, 2019), while sludge production would be reduced by more than 75% by direct anaerobic treatment (Gu *et al.*, 2017). Nitritation-denitritation and mainstream deammonification processes have been considered as a B-stage for nitrogen removal. This chapter first highlights the actions which need to be taken in order to improve energy efficiency in currently adopted processes. In addition, the challenges for process optimization in the A-B processes are elucidated. Finally, other processes for future wastewater treatment plants (WWTPs) and related research needs are outlined.

Author(s): Yu Liu, Jun Gu and Meng Zhang
doi: 10.2166/9781789060089_0147

5.2 OPTIMIZATION OF CURRENT PROCESSES

The CAS-based processes currently adopted in WWTPs are known to be energy intensive and produce a huge amount of waste sludge. Consequently, extensive efforts have been dedicated to optimizing the current process for further improvements in energy efficiency of WWTPs.

5.2.1 Aeration

It is known that aeration-associated energy consumption accounts for about half of the total in-plant energy demand. One way to reduce this is to adopt high efficiency aeration facilities, e.g. fine bubble aerators, aeration control strategies, etc. It has been reported that energy consumption by aeration could be reduced by 10% with fine bubble aeration (Pakenas, 1995). In fact, the aeration capacity for municipal wastewater is mainly determined by the oxygen demand for biological conversion of COD and ammonia in various biological processes. In the CAS-based processes, COD is biologically converted into carbon dioxide and biomass, while nitrogen is removed via nitrification, i.e. oxidizing ammonium to nitrite and nitrate, and denitrification at the cost of high oxygen demand. Therefore, improving aeration efficiency, to some extent, is helpful but would not lead to significant energy saving if total oxygen demand cannot be reduced.

5.2.2 Energy recovery

To improve the overall energy efficiency of WWTPs, waste activated sludge (WAS) is normally subjected to anaerobic digestion (AD) for biogas production and sludge volume reduction, after which the biogas is utilized for energy and heat production in a combined heat and power (CHP) system. Even so, energy efficiency is still as low as 31.0–35.9% (Zhou *et al.*, 2013). To further enhance energy production via AD, co-digestion of WAS and other organic feedstocks (e.g. food waste) has been investigated (Xie *et al.*, 2016; Yin *et al.*, 2016). In this co-digestion process, food waste rich in organics was directly fed into the anaerobic digester with WAS to promote the biogas production. By adding food waste, the generated electric energy through CHP could reach about 180% of the in-plant electricity consumption, with a 5–10% increase in digested solid residues (Nowak *et al.*, 2015). It has been reported that the digested solids residue produced from food waste alone could be used as biofertilizer (Ma & Liu, 2019), while the digested solids residues from co-digestion of WAS and food waste appeared to be unsuitable for agricultural uses due to the safety concerns about heavy metals (Ma *et al.*, 2017). This suggests that the digested solids residues in co-digestion need further treatment via incineration or landfill.

To tackle this issue, an innovative co-digestion of WAS and food waste with enzymatic pretreatment was developed to enhance energy and resource recovery. In this integrated process, food waste is first subjected to enzymatic hydrolysis, followed by solid-liquid separation, after which the produced solids are readily harvested as biofertilizer, while the liquid rich in biodegradable organics is anaerobically digested with WAS, with the purpose of promoting energy recovery and reducing sludge volume. In this process, it was found that solids reduction was about 15% higher that of traditional anaerobic co-digestion of WAS and food waste without pretreatment, whilst the total biomethane production was about four times higher (Ma *et al.*, 2017). This clearly suggests that the energy recovered from this innovative co-digestion process could eventually offset the energy used to drive municipal wastewater treatment, leading to an energy-neutral or even positive operation of WWTPs with the production of value-added biofertilizer.

5.3 ACTIONS FOR MOVING FORWARD WITH A-B PROCESSES

Optimizations for improving in-plant energy efficiency are all based on the currently adopted conventional biological processes. As discussed in Chapters 1 and 2, there is limited room to further improve the performance of the current biological processes which were built on the bio-oxidization philosophy with high energy consumption and the generation of WAS. Indeed, this bio-oxidation-based treatment philosophy is, to some extent, against the core values of the A-B processes. Thus, a paradigm shift in process design appears necessary for municipal wastewater reclamation. With global concerns about climate change and environmental sustainability, more effort into the synergistic integration of A-B processes towards energy neutral or even positive wastewater reclamation is urgently needed.

5.3.1 Selection of A-B processes for municipal wastewater reclamation

In Chapter 3, various configurations of A-B processes are identified for energy-efficient municipal wastewater reclamation. Since the characteristics of municipal wastewater (e.g. sCOD, pCOD, BOD, TN, ammonium, TP, temperature, alkalinity etc.) may vary significantly in different countries, care should be taken in the selection of A-B processes to be used to target wastewater treatment, with considering process performance, process stability, energy recovery, energy consumption and sludge production in a holistic manner. In this regard, guidelines for such process selection should be developed.

5.3.2 Mainstream deammonification

Currently, mainstream deammonification is considered the most suitable B-stage in A-B processes because influent COD can be totally harvested for energy production, with a minimum requirement of aeration for nitrogen removal. However, at this stage, mainstream deammonification processes are still facing some major challenges in terms of sustainable partial nitritation, effective retention of anammox bacteria, etc.

5.3.2.1 Anammox biofilms vs granules in mainstream deammonification

The retention of anammox bacteria is crucial in mainstream deammonification processes. Generally, slow growing anammox bacteria tend to grow in the form of biofilms or granules, while nitrifying sludge preferentially resides in flocs (Laureni *et al.*, 2016; Winkler *et al.*, 2012). In fact, anammox biofilms or granules, and nitrifying flocs usually coexist in the single-stage deammonification process (Lemaire *et al.*, 2014; Lotti *et al.*, 2015). In such a single-stage system, process design and operation strategies for nitrite oxidizing bacteria (NOB) repression are critical, without compromising ammonia oxidizing bacteria (AOB) and anammox activities. It has been reported that a short sludge retention time (SRT) favours NOB washout and the establishment of nitrite shunt (Wang *et al.*, 2019; Xu *et al.*, 2017). Therefore, separating the SRT of suspended nitrifying sludge and fixed-growth anammox bacteria (i.e. biofilms and granules) is one of the effective operation strategies which can be adopted for sustainable mainstream deammonification. Similar to the situation in the sidestream process, differential control of SRT for anammox granules and suspended nitrifying sludge could be realized through biomass selection by cyclones (Wett *et al.*, 2015). However, cyclones cannot retain all anammox bacteria, i.e. small anammox granules could be easily wasted out through overflows from the cyclones (Kwon *et al.*, 2019), potentially leading to system failure. By contrast, anammox biocarriers could be easily retained in the system by screens (Han *et al.*, 2016; Lemaire *et al.*, 2014), indicating that the loss of anammox bacteria could be avoided. More importantly, anammox biocarriers could be readily applied in processes designed with anoxic chambers (Gu *et al.*, 2019b). Therefore, fixed anammox bacteria might offer a better solution for the easy and sustainable operation of the mainstream deammonification process.

5.3.2.2 NOB repression

Sustainable partial nitritation is the prerequisite of mainstream deammonification. As discussed in Chapter 4, the commonly adopted parameters for NOB repression include dissolved oxygen (DO) concentration, SRT, alternating anoxic/oxic conditions, free ammonia (FA) and free nitrous acid (FNA) inhibition, etc. Due to the low concentration of ammonium in municipal wastewater, FA and FNA inhibition are no longer available, while single parameter control appears ineffective for NOB repression; see, for example, the unsuccessful NOB repression with low DO concentration control (Lotti *et al.*, 2015). It has been reported that the combined control of SRT, alternating anoxic/oxic conditions, sludge return ratio and DO concentration appeared to be effective for NOB repression in a pilot-scale step-feed reactor (Wang *et al.*,

2019). This indicates that synergistic control of multi-parameters could offer an engineering solution towards stable mainstream deammonification, while further investigation is definitely needed.

5.3.3 Utilization of dissolved methane for nitrogen removal

The ubiquitous presence of dissolved methane in anaerobic effluent has been considered as a hurdle for the application of anaerobic treatment in municipal wastewater treatment due to concerns for climate change and energy loss (Liu *et al.*, 2014). An alternative to recovering dissolved methane is to utilize it as a carbon source for biological nitrogen removal through nitrite/nitrate-dependent methane oxidation (N-DAMO). It has been shown that the combination of N-DAMO and anammox could achieve a more sustainable nitrogen removal (van Kessel *et al.*, 2018). However, this technology is still facing the challenges of the low growth rate of N-DAMO bacteria, dissolved methane stripping due to normal aeration and competition with heterotrophs, etc.

5.3.4 Emerging technologies

It should be noted that the A-B process represents a new concept for realizing energy-efficient wastewater treatment, with many configurations available. The A-B processes described in this book are relatively more practical with potential applications in the near future. In fact, any emerging technology that can fulfill the requirements could be incorporated into the A-B processes. For example, microbial fuel cell (MFC) technology, illustrated in Chapter 2, is currently challenged by its scalability but in principle it could serve as the A-stage for direct electricity generation, with minimized sludge production.

5.4 PERSPECTIVES

As discussed in Chapters 1 and 2, municipal wastewater can be viewed as a source of water, energy and nutrients which should be synergistically recovered (Liu *et al.*, 2018). However, current processes have been developed on the simple philosophy of removal. Any developments moving forward should be dedicated to shifting any wastewater treatment plant (WWTP) to being a water, power and resources factory, by implementing emerging process configurations such as the A-B process (Figure 5.1).

5.4.1 Energy recovery

A substantial amount of chemical energy is lost along with wastewater treatment in current WWTPs at the cost of high energy consumption. For example, it has been estimated that, in 2012, 16.86×10^9 kWh of electrical energy was consumed in China for municipal wastewater treatment (Liu *et al.*, 2018). However, as shown in Chapter 1, the total recoverable potential energy in municipal wastewater could be estimated at 805×10^9 kWh of electricity per year, at a global scale.

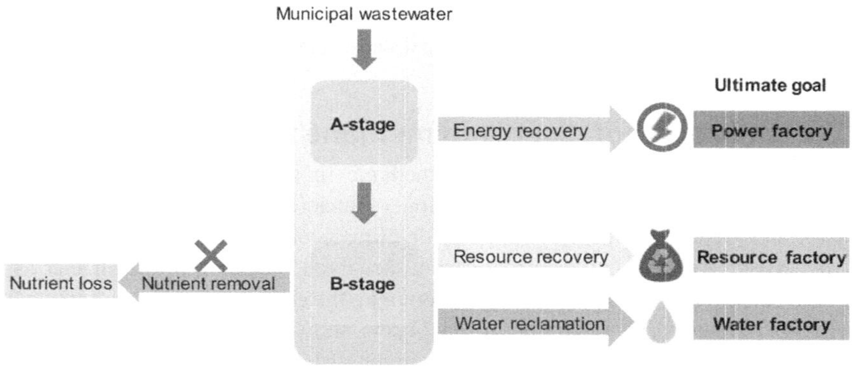

Figure 5.1 Turning a WWTP into a factory of power, resources and water.

Obviously, direct energy recovery from municipal wastewater is beneficial for achieving reduced energy demand, sludge production and GHG emissions. Currently, direct anaerobic treatment has been shown to be effective for energy recovery from municipal wastewater with minimized sludge production (McCarty *et al.*, 2011), while almost all nutrients (e.g. ammonium and phosphate) remained in the anaerobic effluent, allowing further nutrient recovery. An anaerobic membrane bioreactor (AnMBR) is a preferable anaerobic process because of its fast startup (Hu & Stuckey, 2006). Moreover, the solids-free permeate from an AnMBR is suitable for further water reclamation for a high-grade product water (Gu *et al.*, 2019a).

Apart from anaerobic treatment, microbial fuel cells (MFCs) have been investigated for direct energy production from wastewater and hydrolyzed food waste (Ge *et al.*, 2015; Xin *et al.*, 2018). MFC technology could convert the inherent energy in wastewater directly to electricity with significantly lowered sludge production (Zhang *et al.*, 2013). However, MFC application is currently limited by its low power density and it is difficult to scale up.

5.4.2 Wastewater reclamation

It has been reported that municipal wastewater reclamation is the most economically viable strategy when compared with other solutions, e.g. desalination, extraction of underground water, water diversion, etc. (Liu *et al.*, 2018). Therefore, ideally WWTPs should be transformed into a water factory for producing high-grade reclaimed water. For example, with advanced treatment, e.g. reverse osmosis (RO), municipal wastewater could be purified to ultrapure water readily for industrial and indirect potable uses. In current practice, the conventional CAS process coupled with microfiltration (MF) or ultrafiltration (UF) is commonly adopted prior to RO. However, there are some concerns about the relatively high energy consumption and production of WAS associated with the biological process. By contrast, a process integrating AnMBR and RO showed a potential toward producing a high-grade product with a quality comparable to NEWater (Gu *et al.*, 2019a).

5.4.3 Nutrients recovery

As illustrated in Chapter 1, nutrients recovery from municipal wastewater is a necessity for environmental sustainability. Currently, phosphorus is mainly recovered from the sludge digestion liquor in the sidestream in the form of struvite, hydroxyapatite and vivianite (Campos *et al.*, 2018). Similarly, a process known as coupled aerobic-anoxic nitrous decomposition operation (CANDO) has recently been reported in the production of nitrous oxide from sludge digestion liquor (Vasilaki *et al.*, 2019); however, technical feasibility, economic viability and environmental sustainability have not yet fully demonstrated (Zhang *et al.*, 2019). So far, ammonium can be recovered from anaerobic digestion liquor as a fertilizer in the form of ammonium sulphate, via steam stripping, distillation and reversible chemsorption (Zarebska *et al.*, 2015), but energy consumption appears to be significantly high. Clearly, the recovery of nitrogenous compounds (e.g. ammonium, nitrous oxide etc.) and phosphorus from municipal wastewater appears technically challenging and cost ineffective due to the low concentration of nutrients (Liu *et al.*, 2018).

5.5 CONCLUSIONS

Further optimization of the current biological process with the CAS as core might not be an option towards energy neutral or even positive operation. However, the novel A-B processes which have been demonstrated at laboratory and pilot scales have shown a potential for improving the energy efficiency of municipal wastewater reclamation. Moving forward, efforts should be dedicated to developing guidelines for selecting a particular A-B process according to the characteristics of municipal wastewater and target effluent quality. In addition, the challenges of mainstream deammonification (as identified in Chapters 3 and 4) which are the current bottleneck for its large-scale application should also be addressed. The ultimate goal of municipal wastewater reclamation is to concurrently harvest water, power and resources.

REFERENCES

Campos J. L., Crutchik D., Franchi Ó., Pavissich J. P., Belmonte M., Pedrouso A., Mosquera-Corral A. and Val del Río Á. (2018). Nitrogen and phosphorus recovery from anaerobically preteated agro-food wastes: a review. *Frontiers in Sustainable Food Systems*, **2**, 91.

Ge Z., Wu L., Zhang F. and He Z. (2015). Energy extraction from a large-scale microbial fuel cell system treating municipal wastewater. *Journal of Power Sources*, **297**, 260–264.

Gu J., Xu G. and Liu Y. (2017). An integrated AMBBR and IFAS-SBR process for municipal wastewater treatment towards enhanced energy recovery, reduced energy consumption and sludge production. *Water Research*, **110**, 262–269.

Gu J., Yang Q. and Liu Y. (2018). Mainstream anammox in a novel A-2B process for energy-efficient municipal wastewater treatment with minimized sludge production. *Water Research*, **138**, 1–6.

Gu J., Liu H., Wang S., Zhang M. and Liu Y. (2019a). An innovative anaerobic MBR-reverse osmosis-ion exchange process for energy-efficient reclamation of municipal wastewater to NEWater-like product water. *Journal of Cleaner Production*, **230**, 1287–1293.

Gu J., Zhang M., Wang S. and Liu Y. (2019b). Integrated upflow anaerobic fixed-bed and single-stage step-feed process for mainstream deammonification: a step further towards sustainable municipal wastewater reclamation. *Science of The Total Environment*, **678**, 559–564.

Han M., Vlaeminck S. E., Al-Omari A., Wett B., Bott C., Murthy S. and De Clippeleir H. (2016). Uncoupling the solids retention times of flocs and granules in mainstream deammonification: a screen as effective out-selection tool for nitrite oxidizing bacteria. *Bioresource Technology*, **221**, 195–204.

Hu A. Y. and Stuckey D. C. (2006). Treatment of dilute wastewaters using a novel submerged anaerobic membrane bioreactor. *Journal of Environmental Engineering*, **132**(2), 190–198.

Kwon K., Kim H., Kim W. and Lee J. (2019). Efficient nitrogen removal of reject water generated from anaerobic digester treating sewage sludge and livestock manure by combining Anammox and autotrophic sulfur denitrification processes. *Water*, **11**(2), 204.

Laureni M., Falas P., Robin O., Wick A., Weissbrodt D. G., Nielsen J. L., Ternes T. A., Morgenroth E. and Joss A. (2016). Mainstream partial nitritation and anammox: long-term process stability and effluent quality at low temperatures. *Water Research*, **101**, 628–639.

Lemaire R., Zhao H., Thomson C., Christensson M., Piveteau S., Hemmingsen S., Veuillet F., Zozor P. and Ochoa J. (2014). Mainstream deammonification with ANITATMMox process. *Proceedings of the Water Environment Federation*, **10**(6), 2183–2197.

Liu Z. H., Yin H., Dang Z. and Liu Y. (2014). Dissolved methane: a hurdle for anaerobic treatment of municipal wastewater. *Environmental Science & Technology*, **48**(2), 889–890.

Liu Y. J., Gu J. and Liu Y. (2018). Energy self-sufficient biological municipal wastewater reclamation: present status, challenges and solutions forward. *Bioresource Technology*, **269**, 513–519.

Lotti T., Kleerebezem R., Hu Z., Kartal B., de Kreuk M. K., van Erp Taalman Kip C., Kruit J., Hendrickx T. L. and van Loosdrecht M. C. (2015). Pilot-scale evaluation of anammox-based mainstream nitrogen removal from municipal wastewater. *Environmental Technology*, **36**(9–12), 1167–1177.

Ma Y. Q. and Liu Y. (2019). Turning food waste to energy and resources towards a great environmental and economic sustainability: an innovative integrated biological approach. *Biotechnology Advances*. (In press).

Ma Y. Q., Yin Y. and Liu Y. (2017). New insights into co-digestion of activated sludge and food waste: biogas versus biofertilizer. *Bioresource Technology*, **241**, 448–453.

McCarty P. L., Bae J. and Kim J. (2011). Domestic wastewater treatment as a net energy producer: can this be achieved? *Environmental Science & Technology*, **45**(17), 7100–7106.

Nowak O., Enderle P. and Varbanov P. (2015). Ways to optimize the energy balance of municipal wastewater systems: lessons learned from Austrian applications. *Journal of Cleaner Production*, **88**, 125–131.

Pakenas L. J. (1995). Energy Efficiency in Municipal Wastewater Treatment Plants: Technology Assessment. New York State Energy Research and Development Authority, New York, USA.

Rahman A., De Clippeleir H., Thomas W., Jimenez J. A., Wett B., Al-Omari A., Murthy S., Riffat R. and Bott C. (2019). A-stage and high-rate contact-stabilization performance comparison for carbon and nutrient redirection from high-strength municipal wastewater. *Chemical Engineering Journal*, **357**, 737–749.

van Kessel M. A. H. J., Stultiens K., Slegers M. F. W., Guerrero Cruz S., Jetten M. S. M., Kartal B. and Op den Camp H. J. M. (2018). Current perspectives on the application of N-damo and anammox in wastewater treatment. *Current Opinion in Biotechnology*, **50**, 222–227.

Vasilaki V., Massara T. M., Stanchev P., Fatone F. and Katsou E. (2019). A decade of nitrous oxide (N2O) monitoring in full-scale wastewater treatment processes: a critical review. *Water Research*, **161**, 392–412.

Wang H., Xu G., Qiu Z., Zhou Y. and Liu Y. (2019). NOB suppression in pilot-scale mainstream nitritation-denitritation system coupled with MBR for municipal wastewater treatment. *Chemosphere*, **216**, 633–639.

Wett B., Podmirseg S. M., Gomez-Brandon M., Hell M., Nyhuis G., Bott C. and Murthy S. (2015). Expanding DEMON sidestream deammonification technology towards mainstream application. *Water Environmental Research*, **87**(12), 2084–2089.

Winkler M. K., Kleerebezem R. and van Loosdrecht M. C. (2012). Integration of anammox into the aerobic granular sludge process for main stream wastewater treatment at ambient temperatures. *Water Research*, **46**(1), 136–144.

Xie S., Hai F. I., Zhan X., Guo W., Ngo H. H., Price W. E. and Nghiem L. D. (2016). Anaerobic co-digestion: a critical review of mathematical modelling for performance optimization. *Bioresource Technology*, **222**, 498–512.

Xin X., Ma Y. and Liu Y. (2018). Electric energy production from food waste: microbial fuel cells versus anaerobic digestion. *Bioresource Technology*, **255**, 281–287.

Xu G., Wang H., Gu J., Shen N., Qiu Z., Zhou Y. and Liu Y. (2017). A novel A-B process for enhanced biological nutrient removal in municipal wastewater reclamation. *Chemosphere*, **189**, 39–45.

Yin Y., Liu Y. J., Meng S. J., Kiran E. U. and Liu Y. (2016). Enzymatic pretreatment of activated sludge, food waste and their mixture for enhanced bioenergy recovery and waste volume reduction via anaerobic digestion. *Applied energy*, **179**, 1131–1137.

Zarebska A., Romero Nieto D., Christensen K. V., Fjerbæk Søtoft L. and Norddahl B. (2015). Ammonium fertilizers production from manure: a critical review. *Critical Reviews in Environmental Science and Technology*, **45**(14), 1469–1521.

Zhang F., Ge Z., Grimaud J., Hurst J. and He Z. (2013). Long-term performance of liter-scale microbial fuel cells treating primary effluent installed in a municipal wastewater treatment facility. *Environmental Science & Technology*, **47**(9), 4941–4948.

Zhang M., Gu J. and Liu Y. (2019). Engineering feasibility, economic viability and environmental sustainability of energy recovery from nitrous oxide in biological wastewater treatment plant. *Bioresource Technology*, **282**, 514–519.

Zhou Y., Zhang D. Q., Le M. T., Puah A. N. and Ng W. J. (2013). Energy utilization in sewage treatment – a review with comparisons. *Journal of Water and Climate Change*, **4**(1), 1.

Index

IWA Publishing's authorised EU representative for General Product Safety Regulations is Diane D'Arras, 15 rue Duret, 75116 Paris, France, e-mail: safety@iwap.co.uk.

Printed and bound by CPI Group (UK) Ltd, Croydon, CR0 4YY
16/06/2026
02140356-0001